RENEWABLE ENERGY: RESEARCH, DEVELOPMENT AND POLICIES

THE RENEWABLE FUEL STANDARD

OVERVIEW, ISSUES, FUTURE CONSIDERATIONS

RENEWABLE ENERGY: RESEARCH, DEVELOPMENT AND POLICIES

Additional books in this series can be found on Nova's website under the Series tab.

Additional e-books in this series can be found on Nova's website under the e-book tab.

RENEWABLE ENERGY: RESEARCH, DEVELOPMENT AND POLICIES

THE RENEWABLE FUEL STANDARD

OVERVIEW, ISSUES, FUTURE CONSIDERATIONS

LYDIA RIDER
EDITOR

nova publishers
New York

For permission to use material from this book please contact us:
Telephone 631-231-7269; Fax 631-231-8175
Web Site: http://www.novapublishers.com

Library of Congress Cataloging-in-Publication Data

ISBN: 978-1-63321-829-1

Published by Nova Science Publishers, Inc. † New York

CONTENTS

PREFACE

Chapter 1 – Federal policy has played a key role in the emergence of the U.S. biofuels industry. Policy measures have included minimum renewable fuel usage requirements, blending and production tax credits, an import tariff, loans and loan guarantees, and research grants. One of the more prominent forms of federal policy support is the Renewable Fuel Standard (RFS)—whereby a minimum volume of biofuels is to be used in the national transportation fuel supply each year. This report describes the general nature of the RFS mandate and its implementation, and outlines some emerging issues related to the continued growth of U.S. biofuels production needed to fulfill the expanding RFS mandate, the potential inability of the domestic market to absorb ethanol above a 10% share of domestic gasoline fuels (a problem known as the "blend wall"), and the emergence of potential unintended consequences of this rapid expansion.

Congress first established the RFS with the enactment of the Energy Policy Act of 2005 (EPAct, P.L. 109-58). This initial RFS (referred to as RFS1) mandated that a minimum of 4 billion gallons be used in 2006, rising to 7.5 billion gallons by 2012. Two years later, the Energy Independence and Security Act of 2007 (EISA, P.L. 110-140) greatly expanded the biofuel mandate volumes and extended the ramp-up through 2022. The expanded RFS (referred to as RFS2) required the annual use of 9 billion gallons of biofuels in 2008, rising to 36 billion gallons in 2022, with at least 16 billion gallons from cellulosic biofuels, and a cap of 15 billion gallons for corn-starch ethanol.

In addition to the expanded volumes and extended date, RFS2 has three important distinctions from RFS1. First, the total renewable fuel requirement is divided into four separate, but nested categories—total renewable fuels, advanced biofuels, biomass-based diesel, and cellulosic biofuels—each with its own volume requirement. Second, biofuels qualifying under each category must achieve certain minimum thresholds of lifecycle greenhouse gas (GHG) emission reductions, with certain exceptions applicable to existing facilities. Third, all renewable fuel must be made from feedstocks that meet an amended definition of renewable biomass, including certain land use restrictions.

The Environmental Protection Agency (EPA) is responsible for establishing and implementing regulations to ensure that the nation's transportation fuel supply contains the mandated biofuels volumes. EPA's initial regulations for administering RFS1 (issued in April 2007) established detailed compliance standards for fuel suppliers, a tracking system based on renewable identification numbers (RINs) with credit verification and trading, special treatment of small refineries, and general waiver provisions. EPA rules for administering RFS2 (issued in February 2010) built upon the earlier RFS1 regulations and include specific deadlines for announcing annual standards, as well as greater specificity on potential waiver requests and RIN oversight.

Over the long term, the RFS is likely to play a dominant role in the development of the U.S. biofuels sector. However, emerging resource constraints related to the rapid expansion of U.S. corn ethanol production have provoked questions about its long-run sustainability and the possibility of unintended consequences in other markets as well as on the environment. Questions also exist about the ability of the U.S. biofuels industry to meet the expanding mandate for biofuels from non-corn sources such as cellulosic biomass materials, whose production capacity has been slow to develop, or biomass-based biodiesel, which remains expensive to produce owing to the relatively high prices of its feedstocks. Finally, considerable uncertainty remains regarding the development of the infrastructure capacity (e.g., trucks, pipelines, retail pumps, etc.) needed to deliver the expanding biofuels mandate to consumers.

Chapter 2 – Transportation fuels are required by federal law to contain a minimum amount of renewable fuel each year. The renewable fuel standard (RFS), established by the Energy Policy Act of 2005 (EPAct, P.L. 109-58) and amended by the Energy Independence and Security Act of 2007 (EISA, P.L. 110-140), required that a total of 16.55 billion gallons of renewable fuels be used to offset gasoline and diesel fuel in 2013. Under EISA, the scheduled mandates grow each year (to 36 billion gallons in 2022). However, the ability of fuel suppliers to meet the growing RFS mandates has been questioned. For 2014, instead of a total of 18.15 billion gallons scheduled in EISA, the Environmental Protection Agency (EPA) has proposed using waiver authority within the statute to lower the mandate to 15.21 billion gallons. Within the overall RFS there are sub-mandates (or "carveouts") for advanced and cellulosic biofuels. EPA has proposed lowering these as well.

Most of the overall RFS mandate (84% for 2013) is currently met using corn-based ethanol. The vast majority of the ethanol supplied to meet the RFS is blended into gasoline at the 10% level (E10). However, there is a limit to the amount of ethanol that can be blended into gasoline at this level, and other options for supplying ethanol (e.g., higher-level ethanol blends such as E15 and E85) are currently constrained because of infrastructure and other impediments. In prior years this had not been a problem because the volume of biofuels needed to meet the RFS mandate was below the maximum amount of ethanol that could be supplied as E10; but stakeholders were especially concerned about the possibility of facing the "blend wall" in either 2013 or 2014.

The Environmental Protection Agency (EPA) has the authority to waive the RFS requirements, in whole or in part, if certain conditions outlined in the law are present. Under EISA the overall RFS is scheduled to grow to 18.15 billion gallons in 2014. In November 2013 EPA proposed using its general authority to lower the RFS mandates in cases of "inadequate supply." In this case, they interpreted "inadequate supply" to include an inability to *deliver* the necessary biofuels to consumers, even if there is sufficient *production* of said biofuels. Gasoline and diesel fuel suppliers and other stakeholders have generally been supportive of EPA's proposal, while biofuel producers and corn growers believe that EPA is overstepping its authority.

Questions have also been raised over whether the overall mandate diverts enough corn supply from food/feed production to dramatically raise prices in those markets, and whether there is enough feedstock supply and production capacity to meet the carveouts for fuels other than corn ethanol. In 2008 the governor of Texas requested a waiver of the RFS because of high grain prices, although that waiver request was denied because EPA determined that the RFS requirements alone did not "severely harm the economy of a State, a region, or the United States," a standard required by the statute. A similar waiver petition was filed by the governors of several states in August 2012. That petition was denied for similar reasons.

In February 2010, as part of a final rulemaking implementing the RFS as expanded by EISA, EPA waived most of the 2010 cellulosic biofuel carveout—as they have in each subsequent year. In cases where EPA projects inadequate production of cellulosic biofuels to meet the schedule in EISA, the agency must lower the mandate to the projected level, although EPA's process for doing so has been controversial, and the agency's 2012 cellulosic mandate was remanded by the U.S. Court of Appeals for the D.C. Circuit.

This report provides a brief overview of the RFS program and discusses the process and criteria for EPA to waive various parts of the RFS.

Chapter 3 – The federal Renewable Fuel Standard (RFS) was established in the Energy Policy Act of 2005 (EPAct) and significantly expanded in the Energy Independence and Security Act of 2007 (EISA). The RFS requires the use of renewable biofuels in transportation fuel. For 2013, the RFS requires the use of 16.55 billion gallons of renewable fuel. Within the larger mandate, there are sub-mandates ("carve-outs") for advanced biofuels (e.g., biomass-based diesel and cellulosic fuels). By 2022, the RFS requires the use of 36 billion gallons of renewable fuels, including 21 billion gallons of advanced biofuels.

The RFS is a market-based compliance system in which obligated parties (generally refiners and/or terminal operators) must submit credits to cover their obligations. These credits— Renewable Identification Numbers, or RINs—are effectively commodities that can be bought or sold like other commodities. For each gallon of renewable fuel in the RFS program, one RIN is generated. Each RIN is a 38-digit number, with blocks of digits corresponding to various data, including the year the RIN was generated, the producer of the fuel, and the type of fuel. RINs are valid for use in the year they are generated and the following year.

From the beginning of the RFS program, there have been concerns with RIN generation and the RIN market. As the RINs are essentially numbers in a computerized account, there have been errors and opportunities for fraud. Because of concerns over transposed digits, invalid characters, allegations of double-counting (intentional or unintentional) and other errors and inaccuracies, when EPA finalized rules for the RFS as expanded by EISA (the "RFS2"), EPA also established a new in-house trading system in an effort to address these concerns. All RIN transactions must be cleared through this in-house system, called the EPA Moderated Transaction System (EMTS). From the beginning of the RFS2 EPA has maintained that all due diligence remains the duty of obligated parties. Under this "buyer beware" system those purchasing or receiving RINs must certify their validity on their own, and they are responsible for any fraudulent RINs they pass on to other buyers or submit to EPA for compliance.

In late 2011 and early 2012, EPA issued Notices of Violations (NOVs) to three companies that the agency alleges fraudulently generated a combined 140 million biodiesel RINs in 2010 and 2011. Because of these RIN fraud cases, EPA is looking at establishing a system whereby RINs can be certified by third parties registered with EPA. EPA is considering whether such certification would provide obligated parties with an "affirmative defense" if RINs are later found to fraudulent—that is, obligated parties would not be liable for penalties under the Clean Air Act for the use of such RINs. Key questions include whether such an affirmative defense would also eliminate the requirement to purchase make-up RINs. EPA proposed a plan in January 2013, and projects that the rule will be finalized in October 2013.

Most RINs are bought and sold through private contracts registered with the EMTS. However, there are also spot markets for RINs, and since the beginning of 2013, conventional ethanol RIN prices have risen dramatically. Prices rose from roughly $0.07 per gallon in early January to over $1.00 per gallon in early July. Various factors have been identified by stakeholders as potentially causing the price increase, including whether sufficient amounts of ethanol can be blended into gasoline to meet the 2013 RFS mandates and the extent to which non-obligated parties are speculating in RIN markets.

In the 113th Congress several congressional hearings have been held and various bills have been proposed to address both RIN issues and the overall RFS.

Chapter 4 – The Renewable Fuel Standard (RFS) was expanded under the Energy Independence and Security Act of 2007 (EISA; P.L. 110-140) in an effort to reduce dependence on foreign oil, promote biofuel use, and stabilize transportation fuel prices, among other goals. Over 15 years, the RFS requires that increasing amounts of biofuels—36 billion gallons by 2022—be used in transportation fuel. The mandate is to be accomplished in part with advanced biofuels, including cellulosic biofuels—fuels produced from cellulosic materials including grasses, trees, and agricultural and municipal wastes—which would increase over time to comprise some 44% of the RFS in 2022.

The U.S. Environmental Protection Agency (EPA) is required to set the annual standard for cellulosic biofuels under the RFS for the following year by November 30. If projected cellulosic biofuel production is less than the volume specified in the statute, EPA can lower the cellulosic biofuels standard. EPA concluded that the nation lacked sufficient production capacity to meet the RFS cellulosic biofuels mandate for 2010, 2011, 2012, 2013, and 2014. EPA reduced the mandate for 2010 (from 100 million gallons to 5 million gallons actual volume), 2011 (from 250 million gallons to 6.6 million gallons), 2012 (from 500 million gallons to 8.65 million gallons, later vacated by a federal court decision and reduced to zero), and 2013 (from 1 billion gallons to 4 million gallons). EPA proposes to lower the 2014 mandate from

1.75 billion gallons to 17 million ethanol-equivalent gallons, and to rescind the 2011 cellulosic biofuel standard.

The 2010-2012 reduced mandates were not met by actual cellulosic biofuel production, which EPA reports was limited. Instead, these mandates were largely met with waiver credits. EPA reports that the cellulosic biofuels industry is growing incrementally, noting that two commercial-scale cellulosic biofuel facilities began fuel production in 2013, although it is unlikely that enough fuel will be supplied to meet the mandate.

The cellulosic biofuels industry may be able to produce enough fuel to meet the RFS mandates if certain obstacles are overcome: lowering the cost of conversion technology at the initial stages of commercial application, easing access to financing, removing feedstock supply uncertainties, and creating certainty for tax incentives. Another challenge for the cellulosic biofuels industry—and all biofuels industries—is the petroleum industry's opposition to the RFS overall, in part because it views the RFS as unworkable. Other industries—livestock and poultry producers in particular—have joined the petroleum industry in requesting that the RFS be modified, in many cases for reasons unrelated to cellulosic biofuel supply. Another supply constraint is the blend wall—the upper limit to the total amount of ethanol that by law can be blended into U.S. gasoline.

Several federal programs assist the cellulosic biofuels industry, including the U.S. Department of Agriculture's (USDA's) Biorefinery Assistance Program and Biomass Crop Assistance Program, and the U.S. Department of Energy's (DOE's) Loan Guarantee Program. EPA reports that some of the cellulosic biofuel companies identified in its 2014 proposed rule received or were offered significant federal financial support (approximately $387 million) in the form of grants and loan guarantees from USDA and DOE.

Many questions about cellulosic biofuels and the RFS have arisen. Can the RFS mandate for cellulosic biofuels be met? If so, when would it be met? What impact will the continued lowering of the cellulosic biofuels mandate by EPA have on investment in production? Should Congress continue to provide support for cellulosic biofuels, and if so, how? Might Congress statutorily increase the number of qualified feedstocks for the RFS cellulosic biofuels category, given the 112th Congress amendment of the definition of cellulosic biofuels to include algae for some tax incentives? What impact will other legislative discussions (e.g., military support for biofuels) have on the RFS cellulosic biofuels mandate?

This report, in a question and answer format, discusses some challenges facing the cellulosic biofuels community, including feedstock supply estimates, and potential legislative options to address cellulosic biofuels production uncertainty for the RFS.

Chapter 5 – The Renewable Fuel Standard (RFS) establishes minimum volumes of various types of renewable fuels that must be included in the United States' supply of fuel for transportation. Those volumes—as defined by the Energy Independence and Security Act of 2007 (EISA)— are intended to grow each year through 2022. In recent years, the requirements of the RFS have been met largely by blending gasoline with ethanol made from cornstarch. In the future, EISA requires the use of increasingly large amounts of "advanced biofuels," which include diesel made from biomass (such as soybean oil or animal fat), ethanol made from sugarcane, and cellulosic biofuels (made from converting the cellulose in plant materials into fuel).

One of the main goals of the Renewable Fuel Standard is to reduce U.S. emissions of greenhouse gases, which contribute to climate change. EISA requires that the emissions associated with a gallon of renewable fuel be at least a certain percentage lower than the emissions associated with the gasoline or diesel that the renewable fuel replaces. Advanced biofuels and the subcategory of cellulosic biofuels are required to meet more stringent emission standards than those that apply to corn ethanol.

Policymakers and analysts have raised concerns about the RFS, particularly about the feasibility of complying with the standard, whether it will increase prices for food and transportation fuels, and whether it will lead to the intended reductions in greenhouse gas emissions. Because of those concerns, some policymakers have proposed repealing or revising the Renewable Fuel Standard.

In this analysis, the Congressional Budget Office (CBO) evaluates how much the supply of various types of renewable fuels would have to increase over the next several years to comply with the RFS. CBO also examines how food prices, fuel prices, and emissions would vary in an illustrative year, 2017, under three scenarios for the Renewable Fuel Standard:

- *The EISA volumes scenario,* in which fuel suppliers would have to meet the total requirement for renewable fuels, the requirement for advanced biofuels, and the cap on corn ethanol that are stated in EISA for 2017—but not the requirement for cellulosic biofuels, because the capacity to produce enough of those fuels is unlikely to exist by 2017;
- *The 2014 volumes scenario,* in which the Environmental Protection Agency (EPA)—which has some discretion to modify the mandates of

EISA—would keep the RFS requirements for the next several years at the same amounts it has proposed for 2014; and

- *The repeal scenario*, in which lawmakers would immediately abolish the RFS.

The repeal scenario would require Congressional action. In the absence of such action (or of legal restrictions), CBO considers the 2014 volumes scenario much more likely than the EISA volumes scenario, which would require a large and rapid increase in the use of advanced biofuels and would cause the total percentage of ethanol in the nation's gasoline supply to rise to levels that would require significant changes in the infrastructure of fueling stations.

In: The Renewable Fuel Standard
Editor: Lydia Rider
ISBN: 978-1-63321-829-1

Chapter 1

RENEWABLE FUEL STANDARD (RFS): OVERVIEW AND ISSUES*

Randy Schnepf and Brent D. Yacobucci

SUMMARY

Federal policy has played a key role in the emergence of the U.S. biofuels industry. Policy measures have included minimum renewable fuel usage requirements, blending and production tax credits, an import tariff, loans and loan guarantees, and research grants. One of the more prominent forms of federal policy support is the Renewable Fuel Standard (RFS)—whereby a minimum volume of biofuels is to be used in the national transportation fuel supply each year. This report describes the general nature of the RFS mandate and its implementation, and outlines some emerging issues related to the continued growth of U.S. biofuels production needed to fulfill the expanding RFS mandate, the potential inability of the domestic market to absorb ethanol above a 10% share of domestic gasoline fuels (a problem known as the "blend wall"), and the emergence of potential unintended consequences of this rapid expansion.

Congress first established the RFS with the enactment of the Energy Policy Act of 2005 (EPAct, P.L. 109-58). This initial RFS (referred to as RFS1) mandated that a minimum of 4 billion gallons be used in 2006, rising to 7.5 billion gallons by 2012. Two years later, the Energy Independence and Security Act of 2007 (EISA, P.L. 110-140) greatly

* This is an edited, reformatted and augmented version of a Congressional Research Service publication, No. R40155, dated November 22, 2013.

expanded the biofuel mandate volumes and extended the ramp-up through 2022. The expanded RFS (referred to as RFS2) required the annual use of 9 billion gallons of biofuels in 2008, rising to 36 billion gallons in 2022, with at least 16 billion gallons from cellulosic biofuels, and a cap of 15 billion gallons for corn-starch ethanol.

In addition to the expanded volumes and extended date, RFS2 has three important distinctions from RFS1. First, the total renewable fuel requirement is divided into four separate, but nested categories—total renewable fuels, advanced biofuels, biomass-based diesel, and cellulosic biofuels—each with its own volume requirement. Second, biofuels qualifying under each category must achieve certain minimum thresholds of lifecycle greenhouse gas (GHG) emission reductions, with certain exceptions applicable to existing facilities. Third, all renewable fuel must be made from feedstocks that meet an amended definition of renewable biomass, including certain land use restrictions.

The Environmental Protection Agency (EPA) is responsible for establishing and implementing regulations to ensure that the nation's transportation fuel supply contains the mandated biofuels volumes. EPA's initial regulations for administering RFS1 (issued in April 2007) established detailed compliance standards for fuel suppliers, a tracking system based on renewable identification numbers (RINs) with credit verification and trading, special treatment of small refineries, and general waiver provisions. EPA rules for administering RFS2 (issued in February 2010) built upon the earlier RFS1 regulations and include specific deadlines for announcing annual standards, as well as greater specificity on potential waiver requests and RIN oversight.

Over the long term, the RFS is likely to play a dominant role in the development of the U.S. biofuels sector. However, emerging resource constraints related to the rapid expansion of U.S. corn ethanol production have provoked questions about its long-run sustainability and the possibility of unintended consequences in other markets as well as on the environment. Questions also exist about the ability of the U.S. biofuels industry to meet the expanding mandate for biofuels from non-corn sources such as cellulosic biomass materials, whose production capacity has been slow to develop, or biomass-based biodiesel, which remains expensive to produce owing to the relatively high prices of its feedstocks. Finally, considerable uncertainty remains regarding the development of the infrastructure capacity (e.g., trucks, pipelines, retail pumps, etc.) needed to deliver the expanding biofuels mandate to consumers.

INTRODUCTION

Dependence on foreign sources of crude oil, concerns over global climate change, and the desire to promote domestic rural economies have raised interest in renewable biofuels as an alternative to petroleum in the U.S. transportation sector. In response to this interest, U.S. policymakers have enacted a variety of policies, at both the state and federal levels, to directly support U.S. biofuels production and use (although some of these policies have expired).[1] Policy measures have included blending and production tax credits to lower the cost of biofuels to end users, an import tariff to protect domestic ethanol from cheaper foreign-produced ethanol, research grants to stimulate the development of new biofuels technologies, loans and loan guarantees to facilitate the development of biofuels production and distribution infrastructure, and, perhaps most important, minimum usage requirements to guarantee a market for biofuels irrespective of their cost.[2] As a result of expanding policy support, biofuels (primarily corn-based ethanol and biodiesel) production has grown significantly (up over 600%) since the early 2000s. However, despite the rapid growth, U.S. biofuels consumption remains small as a component of U.S. motor fuels, comprising about 5.7% of total transportation fuel consumption (on a gasoline-equivalent basis) in 2012.[3]

Initially, the most significant federal programs for supporting biofuels were tax credits for the production or blending of ethanol and biodiesel into the nation's fuel supply. However, under the Renewable Fuel Standard (RFS)—first established in 2005, then greatly expanded in 2007 (as described below)—Congress mandated biofuels use. In the long term, the expanded RFS usage mandate is likely to prove more significant than tax incentives in promoting the use of these fuels.

This report focuses specifically on the RFS. It describes the general nature of the biofuels RFS and its implementation, and outlines some of the emerging issues related to the sustainability of the continued growth in U.S. biofuels production needed to fulfill the expanding RFS mandate, as well as the emergence of potential unintended consequences of this rapid expansion. This report does not address the broader public policy issue of how best to support U.S. energy policy.

THE RENEWABLE FUEL STANDARD (RFS)

Congress first established a Renewable Fuel Standard (RFS)—a mandatory minimum volume of biofuels to be used in the national transportation fuel supply—in 2005 with the enactment of the Energy Policy Act of 2005 (EPAct, P.L. 109-58). The initial RFS (referred to as RFS1) mandated that a minimum of 4 billion gallons of renewable fuel be used in the nation's gasoline supply in 2006, and that this minimum usage volume rise to 7.5 billion gallons by 2012 (**Table 1**). Two years later, the Energy Independence and Security Act of 2007 (EISA, P.L. 110-140) superseded and greatly expanded the biofuels mandate to 36 billion gallons by 2022. In addition to gasoline, the expanded RFS (referred to as RFS2) applies to most transportation fuel used in the United States—including diesel fuel intended for use in highway motor vehicles, non-road, locomotive, and marine diesel (MVNRLM).[4]

RFS2 directly supports U.S. biofuels production by providing a mandatory market for qualifying biofuels—fuel blenders must incorporate minimum volumes of biofuels in their annual transportation fuel sales irrespective of market prices. By guaranteeing a market for biofuels, RFS2 substantially reduces the risk associated with biofuels production, thus providing an indirect subsidy for capital investment in the construction of biofuels plants. As such, the expanding RFS is expected to continue to stimulate growth of the biofuels industry, particularly for the advanced and cellulosic biofuels categories that are potentially in short supply relative to their growing RFS mandates.

EISA was passed on December 19, 2007, and EPA issued its final rule to implement and administer the RFS2 on February 3, 2010.[5] The new rule builds upon the earlier rule for RFS1. However, there are four major distinctions between RFS1 and RFS2:

- First and foremost, RFS2 increases the mandated usage volumes and extends the time frame over which the volumes ramp up through at least 2022 (**Table 1**).
- Second, RFS2 subdivides the total renewable fuel requirement into four separate but nested categories—total renewable fuels, advanced biofuels, biomass-based diesel, and cellulosic biofuels—each with its own volume requirement or standard (described below).
- Third, biofuels qualifying under each nested RFS2 category must achieve certain minimum thresholds of lifecycle greenhouse gas

(GHG) emission performance, with exceptions applicable to facilities existing or under construction when EISA was enacted (**Table 2**).[6]

- Fourth, under RFS2 all renewable fuel must be made from feedstocks that meet a revised definition of renewable biomass, including certain land use restrictions.[7]

The RFS is administered by the Environmental Protection Agency (EPA).[8] EPA issued its final rule for administering RFS1 in April 2007.[9] This rule established detailed compliance standards for fuel suppliers, a tracking system based on renewable identification numbers (RINs) with credit verification and trading, provisions for treatment of small refineries, and general waiver provisions. EPA rules for administering RFS2 (issued in February 2010) built upon the earlier RFS1 regulations and include specific deadlines for announcing annual standards, as well as greater specificity on potential waiver requests and RIN oversight.

Table 1. EISA 2007 Expansion of the Renewable Fuel Standard (in billions of gallons)

Year	RFS1 Biofuel mandate in EPAct of 2005	RFS2 biofuel mandate				
		Total Renewable fuels	Cap on corn starch-derived ethanol	Portion to be from advanced biofuels		
				Total advanced[h]	Cellulosic	Biomass-based diesel
2006	4.0	—	—	—	—	—
2007	4.7	—	—	—	—	—
2008	5.4	9.00	9.0	0.00	0.00	0.00
2009	6.1	11.10	10.5	0.60	0.00	0.00
2010	6.8	12.95	12.0	0.95	0.0065[a]	1.15b
2011	7.4	13.95	12.6	1.35	0.006	0.80
					(0.00)[c]	
2012	7.5	15.20	13.2	2.00	0.00[d]	1.00
2013	7.6 (est.)	16.55	13.8	2.75	0.006[e]	1.28[e]
2014	7.7 (est.)	18.15	14.4	3.75	1.75	1.28 [h]
		(15.21)[i]	(13.0)[h]	(2.20)[h]	(0.017)[h]	
2015	7.8 (est.)	20.50	15.0	5.50	3.00	[f]
2016	7.9 (est.)	22.25	15.0	7.25	4.25	[f]
2017	8.1 (est.)	24.00	15.0	9.00	5.50	[f]
2018	8.2 (est.)	26.00	15.0	11.00	7.00	[f]
2019	8.3 (est.)	28.00	15.0	13.00	8.50	[f]
2020	8.4 (est.)	30.00	15.0	15.00	10.50	[f]

Table 1. (Continued)

Year	RFS1 Biofuel mandate in EPAct of 2005	RFS2 biofuel mandate				
		Total Renewable fuels	Cap on corn starch-derived ethanol	Portion to be from advanced biofuels		
				Total advanced[h]	Cellulosic	Biomass-based diesel
2021	8.5 (est.)	33.00	15.0	18.00	13.50	[f]
2022	8.6 (est.)	36.00	15.0	21.00	16.00	[f]
2023	—	[g]	[g]	[g]	[g]	[g]

Source: RFS1 is from EPAct (P.L. 109-58), Section 1501; RFS2 is from EISA (P.L. 110-140), Section 202.

[a] The initial EISA cellulosic biofuels mandate for 2010 was for 100 million gallons. On February 3, 2010, EPA revised this mandate downward to 6.5 million ethanol-equivalent gallons. It should be noted that the definition of "cellulosic biofuel" was changed between the RFS1 and RFS2 and that some qualifying fuel produced in the first phase of the program no longer qualifies as cellulosic biofuel.

[b] The biomass-based diesel mandate for 2010 combines the original EISA mandate of 0.65 billion gallons (bgals) with the 2009 mandate of 0.5 bgals.

[c] The initial RFS for cellulosic biofuels for 2011 was 250 million gallons. In November 2010 EPA revised this mandate downward to 6.0 million ethanol-equivalent gallons. In its preliminary proposal for 2014, EPA proposed rescinding the 2011 standard to reflect a lack of supply in 2011 and to respond to the issues raised in the appeals court ruling of January 2013 (see next table note).

[d] The initial RFS for cellulosic biofuels for 2012 was 500 million gallons. In December 2011 EPA revised this mandate downward to 10.45 million ethanol-equivalent gallons. In January 2013, the U.S. Court of Appeals for D.C. vacated EPA's initial cellulosic mandate for 2012 and remanded EPA to replace it with a revised mandate. On February 28, 2013, EPA dropped the 2012 RFS for cellulosic biofuels to zero.

[e] The initial 2013 cellulosic RFS was 1 bgals. In January 2013, EPA preliminarily revised this mandate to 14 million ethanol-equivalent gals.; then in August 2013, finalized it at 6 million ethanol-equivalent gals. The 2013 biodiesel mandate was revised upwards from 1 bgals to 1.28 bgals actual volume.

[f] To be determined by EPA through a future rulemaking, but no less than 1.0 billion gallons.

[g] To be determined by EPA through a future rulemaking.

[h] Advanced biofuels explicitly excludes corn-starch ethanol from inclusion. The preliminary 2014 RFS proposed by EPA (released on November 15, 2013) is in parentheses. Once the proposal is published in the *Federal Register*, it will be open to a 60-day comment period.

[i] Total non-corn-starch or advanced biofuels includes both cellulosic and biomass-based diesel biofuels as well as other qualifying biofuels such as imported sugar-cane-based ethanol.

Four Biofuel Categories

RFS2 includes four biofuel categories, each with a specific volume mandate and lifecycle GHG emission reduction threshold (as compared to the lifecycle GHG emissions of the 2005 baseline average gasoline or diesel fuel that it replaces). Each is also subject to biomass feedstock criteria.

- **Total renewable fuels**. The scheduled mandate grows from nearly 13 billion gallons (bgals) in 2010 to 36 bgals in 2022. Biofuels from new facilities must reduce lifecycle GHG emissions by at least 20% relative to conventional fuels to qualify as a renewable fuel. Most biofuels, including corn-starch ethanol from new facilities, qualify for this mandate. However, the volume of corn-starch ethanol included in the RFS is capped at 13.8 bgals in 2013, but grows to 15 bgals by 2015 and is fixed thereafter.
- **Advanced biofuels.**[10] The mandate grows from nearly 1 bgals in 2010 to 21 bgals in 2022. Advanced biofuels must reduce lifecycle GHG emissions by at least 50% to qualify. A subcomponent of the total renewable fuels mandate, this category includes biofuels produced from non-corn feedstocks—corn-starch ethanol is expressly excluded from this category. Potential feedstock sources include grains such as sorghum and wheat. Imported Brazilian sugarcane ethanol, as well as biomass-based biodiesel and biofuels from cellulosic materials (including non-starch parts of the corn plant such as the stalk and cob), also qualify. The total advanced biofuel mandate for 2013 is 2.75 bgals (ethanol equivalent).
- **Cellulosic and agricultural waste-based biofuel**. The mandate grows from 100 million gallons in 2010 to 16 bgals in 2022 (subsequently, RFS mandates were lowered for 2010, 2011, 2012, and 2013—see discussion under "Waivers to Annual Biofuel Standards"). Cellulosic biofuels must reduce lifecycle GHG emissions by at least 60% to qualify. Cellulosic biofuels are renewable fuels derived from cellulose, hemicellulose, or lignin. This includes cellulosic biomass

ethanol as well as any biomass-to-liquid fuel such as cellulosic gasoline or diesel.

- **Biomass-based biodiesel (BBD)**. The mandate grows from 0.5 bgals in 2009 to 1 bgals in 2012.[11] Any diesel fuel made from biomass feedstocks (including algae) qualifies, including biodiesel (mono-alkyl esters) and non-ester renewable diesel (e.g., cellulosic diesel).[12] The lifecycle GHG emissions reduction threshold is 50%. EPA established the 2013 mandate at 1.28 bgals (actual volume). EPA is proposing the same 1.28 bgals (actual volume) for 2014. BBD produced from cellulosic feedstocks could potentially be used to simultaneously meet the cellulosic and BBD standards.

Usage Volume Requirements

RFS2 is essentially a biofuels mandate with limits on corn-starch ethanol inclusion and carve-outs for higher-performing biofuels (**Figure 1**)—as measured by reductions in lifecycle GHG emissions. The cap on the volume of ethanol derived from corn starch that can be counted under the RFS is intended to encourage the use of non-corn-based biofuels. As a result, corn-starch ethanol blended in excess of its annual cap is not credited toward the annual total RFS.

Nested Categories

Because of the nested nature of the biofuel categories, any renewable fuel that meets the requirement for cellulosic biofuels or biomass-based diesel (BBD) is also valid for meeting the advanced biofuels requirement. Thus, if any combination of cellulosic biofuels or BBD were to exceed their individual mandates, the surplus volume would count against the advanced biofuels mandate, thereby reducing the potential need for imported sugar-cane ethanol or other fuels to meet the unspecified portion of the advanced biofuels mandate (which grows to 21 bgals by 2022).[13]

Similarly, any renewable fuel that meets the requirement for advanced biofuels is also valid for meeting the overall total renewable fuel requirement (which grows to 36 bgals by 2022). As a result, any combination of cellulosic biofuels, BBD, or imported sugar-cane ethanol that exceeds the advanced biofuel mandate would reduce the potential need for corn-starch ethanol to meet the overall mandate.

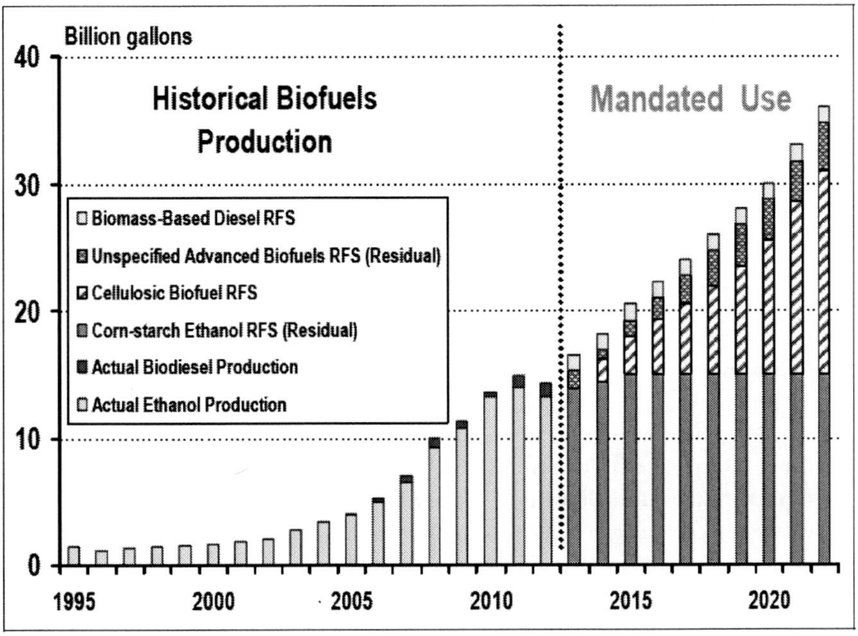

Source: Actual ethanol and biodiesel production data for 1995-2012 are from the U.S. Energy Information Administration (EIA), Department of Energy; the RFS2 mandates by category for 2013-2022 are from EISA (P.L. 110-140).

Figure 1. Renewable Fuels Standard (RFS2) vs. U.S. Biofuel Production Since 1995.

Required Reduction in Lifecycle Greenhouse Gas Emissions

In addition to volume mandates, EISA specified that the lifecycle GHG emissions of a qualifying renewable fuel must be less than the lifecycle GHG emissions of the 2005 baseline average gasoline or diesel fuel that it replaces.[14] EISA established lifecycle GHG emission thresholds for each of the RFS2 biofuels categories (**Table 2**). With respect to the GHG emissions assessments, EISA specifically directed EPA to evaluate the aggregate quantity of GHG emissions (including direct emissions and significant indirect emissions, such as significant emissions from land use changes) related to the full lifecycle, including all stages of fuel and feedstock production, distribution, and use by the ultimate consumer.

Table 2. EISA-Mandated Reductions in Lifecycle GHG Emissions by Biofuel Category (percent reduction from 2005 baseline for gasoline or diesel fuel)

Biofuels category	Threshold reduction
Renewable fuel[a]	20%
Advanced biofuels	50%
Biomass-based diesel	50%
Cellulosic biofuel	60%

Source: "Regulatory Announcement: EPA Finalizes Regulations for the National Renewable Fuel Standard Program for 2010," EPA-420-F-10-007, Office of Transportation and Air Quality, EPA, February 3, 2010.

[a]. The 20% criterion applies to renewable fuel from facilities that commenced construction after December 19, 2007, the date EISA was signed into law. EISA further exempted facilities that operate on natural gas through 2009. Plants that existed or commenced construction prior to these dates are grandfathered in, and are not subject to any GHG emissions requirement.

Fuel Pathways (including ILUC) Meeting Lifecycle GHG Thresholds

Under EISA, EPA is required to evaluate the lifecycle emissions of all biofuel pathways registered in the RFS2. A key point of contention was the inclusion in the statute of a requirement that EPA incorporate so-called "indirect land use changes" (ILUC) in the GHG emissions assessment.[15] ILUC refers to the idea that diversion of an acre of traditional field cropland in the United States to grow a biofuels feedstock crop might result (due to market price effects) in that same acre reappearing at another location and potentially on virgin soils, such as the Amazon rainforest. Such a transfer—when included in the lifecycle GHG calculation of a particular biofuel—could result in an estimated net increase in GHG emissions. EPA's initial assessment of biofuel lifecycle GHG emissions in its proposed RFS2 was met with criticism from many stakeholders.

Several environmental and academic groups argued that, as a result of ILUC costs, corn ethanol should not be permissible under the RFS2. Biofuels proponents argued that ILUC was too vague a concept to be measurable in a meaningful way, and that it alone should not determine the fate of the U.S. biofuels industry.

As a result, EPA reconsidered all of the evidence (including ILUC) and made relevant adjustments to its analytical tools. The resultant changes were announced as part of its final RFS2 rule of February 3, 2010.[16]

In addition, EPA has pointed out that other pathways are likely to be similar enough to existing qualifying pathways (**Table 3**) that they can be extended the same GHG reduction compliance determinations.[17] However, EPA stated that, although the announced determinations for the qualifying fuel pathways (**Table 3**) are final for the time being, its lifecycle methodology remains subject to new developments in the state of scientific knowledge, and that future reassessments may alter the current status of these fuel pathways.

EPA says that it will be able to make determinations on several other potential biomass crops and their fuel pathways in the future.[18] For example, in a February 2013 rule that qualified several new fuel pathways for cellulosic biofuel production, EPA stated that it hoped to provide opportunities to increase the volume of advanced, low-GHG renewable fuels.[19] For other biofuel pathways not yet modeled, EPA encourages parties to use a petition process to request EPA to examine additional pathways.

Table 3. Qualifying Fuel Pathways for Lifecycle GHG Emissions by Biofuel Category

Renewable Fuel—20% GHG Reduction
• Biofuel from the capacity of plants or production facilities that commenced construction after December 19, 2007, unless subject to more stringent criteria as listed below.[a] • Ethanol produced from corn starch at a new natural gas-fired facility (or expanded capacity from an existing facility) using advanced efficient technologies. • Biobutanol from corn starch. • Ethanol made from barley or grain sorghum at dry mill facilities that use natural gas for process energy. • Note: Biodiesel and renewable diesel produced from palm oil do not meet the minimum 20% lifecycle GHG reduction threshold needed to qualify as renewable fuel under the RFS program.
Advanced biofuels—50% GHG Reduction
• Ethanol produced from sugarcane (as in Brazil). • Naphtha and liquefied petroleum gas (LPG) from camelina oil. • Ethanol made from barley or grain sorghum at dry mill facilities that use specified forms of biogas for both process energy and most electricity production.

Table 3. (Continued)

Biomass-based diesel (including jet fuel and heating oil)—50% GHG Reduction
• Biodiesel and non-ester renewable diesel from soy oil, non-food grade corn oil, camelina oil, algal oils, waste oils, fats, and greases. • Biodiesel produced using esterification (a new process method) from soybean oil, oil from annual cover crops, algal oil, biogenic waste oils, fats, and greases, non-food grade corn oil, Canola or rapeseed oil, and camelina oil. • Non-ester renewable diesel based on electricity or natural gas for process energy and feedstocks of soybean oil, oil from annual cover crops, algal oil, biogenic waste oils, fats, and greases, or the non-cellulosic portions of food wastes (i.e., non-food grade corn oil, and camelina oil). • Biodiesel produced using a glycerolysis production process element combined with the traditional transesterification process from free fatty acids (FFA).
Cellulosic biofuels (either cellulosic ethanol or cellulosic diesel)—60% GHG Reduction
• Cellulosic biofuels based on perennial grasses including switchgrass, miscanthus, energy cane, giant reed, and napier grass. • Cellulosic biofuels produced from crop residue (e.g., corn stover, wheat straw, rice straw, and citrus residue), forest material (including slash, pre-commercial thinnings, and solid tree residue remaining from forest product production), secondary annual cover crops planted on existing crop land (e.g., winter cover crops), cellulosic components of separated food and yard waste (including biogenic waste from food processing), and separated municipal solid waste using certain processes identified below. • The following processes—all utilizing natural gas, biogas, and/or biomass as the only process energy sources— qualify as cellulosic biofuel: thermochemical pyrolysis; thermochemical gasification; biochemical direct fermentation; biochemical fermentation with catalytic upgrading; and any other process that uses biogas and/or biomass as the only process energy sources.

Source: EPA announcements of various rules and determinations as posted at "Renewable Fuels: Regulations and Standards," at http://www.epa.gov/otaq /fuels/renewablefuels/regulations.htm.

Notes: This table is not an official, inclusive listing of EPA certified biofuel pathways. For information on potential new biofuel feedstock or production process pathways currently under review, as well as completed pathways, see EPA, "Guidance on New Fuel Pathway Approval Process" at

http://www.epa.gov/otaq/fuels/renewablefuels/compliancehelp/rfs2-lca-pathways.htm.

a. Biofuels from plants that existed or commenced construction prior to this date are grandfathered in, and are not subject to any GHG emissions requirement. Further, the grandfathering provisions extend through 2009 for plants that use natural gas.

Grandfathered Plants

Fuel from the capacity of facilities that either existed or commenced construction prior to December 19, 2007 (the date of enactment of EISA), is exempt from the 20% lifecycle GHG threshold requirement. The exemption is extended to ethanol facilities that commenced construction on or before December 31, 2009, provided that those facilities use natural gas, biofuels, or a combination thereof as processing fuel. However, any new expansion of production capacity at existing facilities must be designed to achieve the 20% GHG reduction threshold if the facility wants to generate RINs for that volume.

Feedstock Requirements

EISA changed the definition of renewable fuel to require that it be made from feedstocks that qualify under an amended definition of "renewable biomass."[20] As such, EISA limits not only the types of feedstocks that can be used to make renewable fuel, but also the land that these renewable fuel feedstocks may come from. Specifically excluded under the EISA definition are virgin agricultural land cleared or cultivated after December 19, 2007, as well as tree crops, tree residues, and other biomass materials obtained from federal lands. These restrictions are applicable to both domestic and foreign feedstock and biofuels producers.

Existing agricultural land includes three land categories—cropland, pastureland, and Conservation Reserve Program (CRP) land. Rangeland is excluded. Fallow land is defined as idled cropland and is therefore included within the definition of agricultural land.

EPA determined that fuels produced from five categories of feedstocks (primarily targeted for cellulosic biofuels) were expected to have less or no indirect land use change and thereby qualify as renewable biomass:[21]

- crop residues such as corn stover, wheat straw, rice straw, citrus residue;

- forest material including eligible forest thinnings and solid residue remaining from forest product production;
- secondary annual crops planted on existing cropland, such as winter cover crops;
- separated food and yard waste, including biogenic waste from food processing; and
- perennial grasses, including switchgrass and miscanthus.

Implementation of the RFS

The EPA is responsible for revising and implementing regulations to ensure that the national transportation fuel supply sold in the United States during a given year contains the mandated volume of renewable fuel in accordance with the four nested volume mandates of the RFS2.[22] To accomplish this task, EPA first calculates annual percentage standards for the four biofuel categories of RFS2. The percentage standards apply to refiners, blenders, and importers of gasoline and diesel fuels and are used to determine each individual company's renewable volume obligation (RVO). To facilitate meeting the requirements, while taking into consideration regional differences in biofuels production and availability, EPA established a system of tradable RINs. Percentage standards, RVOs, and RINs are described in this section.

Determining Annual Percentage Standards

In order to ensure that the requisite volumes of biofuels are used each year, EPA first estimates the total volume of gasoline and diesel fuel that is expected to be used in the United States during the upcoming year. EPA relies on projections from the Department of Energy's Energy Information Agency (EIA) for this estimate.[23] The percentage obligation (or standard) is computed as the total amount of renewable fuels mandated to be used in a given year expressed as a percentage of expected total U.S. transportation fuel use (**Table 4**). This ratio is adjusted to account for the small refinery exemptions. A separate ratio is calculated for each of the four biofuel categories.

Under EISA, EPA is required to set the biofuel standards on a final basis by November 30 for the following year, based in part on information provided by the EIA. In order to accommodate this deadline, EPA announced that it intended to issue a notice of proposed rulemaking (NPRM) by summer of the preceding year, and on a final basis by November 30 of the preceding year.[24] These announcements are to include the cellulosic biofuel waiver credit price (see section on "Cellulosic Biofuel Waiver Credits") and the status of the aggregate compliance approach to land-use restrictions under the definition of renewable biomass for both the United States and Canada.

Table 4. RFS Standards: 2013 Final versus 2014 Proposed

RFS Category	Percentage Ratio (%)		Volume of Renewable Fuel (billion gallons)		
	2013 Final	2014 Proposed	2013 Final	2014 Proposed	2014 (EISA)[a]
Cellulosic biofuels	0.004%	0.010%	0.006	0.017	1.75
Biomass-based diesel	1.13%	1.16%	1.28	1.28	1.28[b]
Advanced biofuels	1.62%	1.33%	2.75	2.20	3.75
Total renewable fuel	9.74%	9.20%	16.55	15.21	18.15
Implied corn ethanol cap[c]	—	—	13.8	13.01	14.40

Source: EPA Fact Sheet, "EPA Finalizes 2013 Renewable Fuel Standards," *EPA-420-F-13-042,* August 2013; and "EPA Proposes 2014 Renewable Fuel Standards, 2015 Biomass-Based Diesel Volumes," *EPA-420-F-13-048*, November 2013.

Notes: All volumes are given in ethanol-equivalent gallons except for biomass-based diesel, which is given in terms of physical volume.

[a] Original RFS as scheduled in EISA (P.L. 110-140).

[b] See foonote h in Table 1.

[c] The implied corn ethanol cap under the RFS is the difference between the total and the advanced categories.

Equivalence Values

The equivalence value (EV) of a renewable fuel represents the number of gallons that can be claimed for compliance purposes for every physical gallon of renewable fuel. Under RFS1, the EV was based on the energy content of each renewable fuel relative to ethanol. As a result, the EV for ethanol was 1.0; butanol was 1.3; biodiesel (mono-alkyl ester) was 1.5; and non-ester renewable diesel was 1.7. Cellulosic biofuel was granted a 2.5-to-1 credit. In general, these equivalence values were continued in the RFS2, with one key exception. For purposes of meeting the biomass-based biodiesel standard, each gallon of BBD will count as 1.0; however, for purposes of meeting the advanced biofuel standard, cellulosic standard and/or the total renewable biofuel standard, each gallon of BBD will count as 1.5 or 1.7 (depending on the type of fuel) in order to reflect its higher energy content. Under the RFS2, the 2.5-to-1 bonus for cellulosic biofuel was eliminated.

Determining an Individual Company' s Obligation

The RFS mandates (by biofuel category) are ultimately enforced on gasoline and diesel fuel refiners, blenders, and/or importers (not on biofuels producers or importers). Companies that supply gasoline or diesel transportation fuel for the retail market are obligated to include a quantity of biofuels equal to a percentage of their total annual fuel sales—referred to as a renewable volume obligation (RVO). The RVO is obtained by applying the EPA-announced standards for each of the four biofuel categories to the firm's annual fuel sales to compute the mandated biofuels volume. At the end of the year, each supplier must have enough RINs to show that it has met its share of each of the four mandated standards. Failure to acquire sufficient RINs to meet a party's RVO (see section "Renewable Identification Numbers (RINs)" for details) is subject to civil penalties of up to $32,500 per day, plus the amount of any economic benefit or savings resulting from the violation.[25]

Since the RFS program is intended to require a specific volume of renewable fuel to be consumed in the U.S. domestic transportation fuel market, RINs associated with exports of renewable fuel are not valid for RFS compliance purposes. To ensure that renewable fuels exported from the U.S. cannot be used by an obligated party for RFS compliance purposes, the RINs associated with any exported renewable fuel must be removed from circulation. To achieve this under the RIN-based program, exporters of renewable fuel are assigned an RVO that is equal to the annual volume of renewable fuel that they export.

Waivers to Annual Biofuel Standards

EISA requires that EPA evaluate and make an appropriate market determination for setting the RFS standards each year. As part of this process, EPA announced that it will issue a notice of proposed rulemaking by summer and a final rule by November 30 of each year to set the renewable fuel standard for each ensuing year.[26] Pursuant to this task, the EPA Administrator has the authority to waive the RFS requirements, in whole or in part, if, in her determination, there is inadequate domestic supply to meet the mandate, or if "implementation of the requirement would severely harm the economy or environment of a State, a region, or the United States."[27]

In addition to waivers associated with the annual RFS review process, EPA may respond to waiver requests resulting from unusual circumstances. For example, in 2008 the governor of Texas requested a waiver of the RFS because of high grain prices; however, that waiver request was denied because EPA determined that the RFS requirements alone did not "severely harm the economy ... of a State, a region, or the United States," a standard required by the statute. A similar waiver was requested in 2012 by the governors of Arkansas and North Carolina and several other states.[28] In both cases, the petition was ultimately denied.[29]

Under certain conditions, the EPA administrator may waive (in whole or in part) the specific carve-outs for cellulosic biofuel and biomass-based diesel fuel. Because these categories are nested, EPA may pass the waiver along to the higher, aggregate totals for advanced and total renewable biofuels. However, through 2013 EPA has elected to limit the waiver to the cellulosic category. For example, in each of the years 2010 through 2013 EPA has waived the original RFS mandates for cellulosic biofuels, but left both the advanced and total renewable mandates unchanged as follows:

- In February 2010, EPA lowered the 2010 RFS for cellulosic biofuels to 6.5 million gallons (mgals), on an ethanol-equivalent basis, down from its original 100 mgals scheduled by EISA.[30]
- In November 2010, EPA lowered the 2011 RFS for cellulosic biofuels to 6 mgals (ethanol equivalent), down from its original 250 mgals.[31]
- In December 2011, EPA lowered the 2012 RFS for cellulosic biofuels to 8.65 mgals (ethanol equivalent), down from its original 500 mgals.[32]
- In August 2013, EPA lowered the 2013 RFS for cellulosic biofuels to 6 mgals (ethanol equivalent), down from its original 1 billion gallons.[33]

EPA cited a lack of current and expected production capacity, driven largely by a lack of investment in commercial-scale refineries—for example, only a limited number of cellulosic biofuel RINs were registered in 2012 (20,069 gallons), while no commercial production was reported in 2010 and 2011.[34] The downward revisions for 2010 through 2013 suggest that the actual cellulosic biofuels standard for future years, although explicitly scheduled in statute, is uncertain.

In contrast to these previous waivers, which were limited to the cellulosic biofuels mandate, on November 15, 2013, EPA released its proposed 2014 RFS mandates and included a waiver of cellulosic, advanced, and total renewable biofuels. EPA is proposing to use both its cellulosic waiver authority and its general authority to lower the total 2014 RFS below the 2013 levels. EPA has determined that the current "blend wall" limitation of 10% ethanol in gasoline[35] would lead to inadequate supply of RINs to comply with the mandate.[36] Therefore, EPA is proposing lowering the advanced biofuel mandate by 1.55 billion gallons,[37] and the overall RFS by 2.94 billion gallons,[38] from the amounts scheduled for 2014 in the statute. Under the proposal, the corn-ethanol cap would be roughly 1.4 billion gallons below the level scheduled in the statute for 2014, and roughly 800 million below the 2013 cap. Some ethanol proponents argue that EPA has exceeded its legislative waiver authority and that this action will likely result in litigation.[39]

Cellulosic Biofuel Waiver Credits

If EPA reduces the required volume of cellulosic biofuel according to the waiver provisions in EISA, EPA must offer a number of credits to obligated parties no greater than the reduced cellulosic biofuel standard. These waiver credits are not allowed to be traded or banked for future use, and are only allowed to be used to meet the cellulosic biofuel standard for the year that they are offered. The formula for determining the value of the credits is set in statute. Since the cellulosic standard was lowered for each year from 2010 through 2013, cellulosic waiver credits were made available to obligated parties at announced prices per gallon—$1.56 in 2010; $1.13 in 2011; $0.78 in 2012; and $0.42 in 2013. The value of the credits is equal to the amount by which $3.00 per gallon—adjusted for inflation—exceeds the average wholesale price of a gallon of gasoline in the United States in the preceding year.

Unachieved Cellulosic Biofuels Mandates

In 2012 the American Petroleum Institute (API) challenged the obligation under the RFS to use cellulosic biofuels that do not exist in sufficient amounts in commercial markets or pay a fee. After three successive years (2010-2012) where, first, EPA lowered the cellulosic biofuels mandate, cellulosic biofuels production failed to achieve the lowered mandates. API petitioned the U.S. Court of Appeals, D.C., charging that EPA exceeded its authority by setting unachievable standards in an effort to promote cellulosic biofuel development. On January 25, 2013, the appeals court agreed with API's charge, ruling that the EPA's cellulosic biofuels mandate for 2012 was vacated and that EPA must replace it with a revised mandate. On February 27, 2013, EPA announced that the 2012 cellulosic biofuel standard was vacated (dropped to zero).[40] Then, in November, 2013, EPA proposed retroactively lowering the 2011 RFS to zero.[41]

Renewable Identification Numbers (RINs)

A RIN is a unique 38-character number that is issued (in accordance with EPA guidelines) by the biofuel producer or importer at the point of biofuel production or the port of importation.[42] Each qualifying gallon of renewable fuel has its own unique RIN. RINs are generally assigned by batches of renewable fuel production as follows:

RIN = KYYYYCCCCFFFFFBBBBBRRDSSSSSSSSEEEEEEEE

Where

K= code distinguishing RINs still assigned to a gallon from RINs already detached

YYYY = the calendar year of production or import

CCCC = the company ID

FFFFF = the company plant or facility ID

BBBBB = the batch number

RR = the biofuel equivalence value (described below)

D = the renewable fuel category

SSSSSSSS = the start number for this batch of biofuel

EEEEEEEE = the end number for this batch of biofuel

Under the RFS2 RIN formulation, Code D has been redefined to identify which of the four RFS categories—total, advanced, cellulosic, or biodiesel—the biofuel satisfies (**Table 5**).

Table 5. RFS D Code Definitions

D value	RFS1	RFS2
1	Cellulosic biomass ethanol	na
2	Any other renewable fuel	na
3	na	Cellulosic biofuel
4	na	Biomass-based diesel
5	na	Advanced biofuel
6	na	Renewable fuel
7	na	Cellulosic diesel

Source: EPA, 40 C.F.R. Part 80, "Regulation of Fuels and Fuel Additives: Changes to Renewable Fuel Standard Program, Final Rule," Feb. 3, 2010.

Notes: na = not applicable.

Together, SSSSSSSS and EEEEEEEE identify the RIN block that demarcates the number of gallons of renewable fuel the batch represents in the context of compliance with the RFS—that is, RIN gallons. The RIN-gallon total equals the product of the liquid volume of renewable fuel times its equivalence value. For example, since biodiesel has an equivalence value of 1.5 when being used as an advanced biofuel, 1,000 gallons of biodiesel would equal 1,500 RIN gallons of advanced biofuels. If the RIN block start for that batch was 1 (i.e., SSSSSSSS = 00000001), then the end value (EEEEEEEE) would be 00001500, and the RR code would be RR = 15.

Any party that owns RINs at any point during the year (including domestic and foreign producers, refiners, exporters, and importers of renewable fuels) must register with the EPA and follow RIN record-keeping and reporting guidelines. RINs can only be generated if it can be established that the feedstock from which the fuel was made meets EISA's definitions of renewable biomass, including land restrictions. The feedstock affirmation and record-keeping requirements apply to RINs generated by both domestic renewable fuel producers and RIN-generating foreign renewable fuel producers or importers. After a RIN is created by a biofuel producer or importer, it must be reported to the EPA. When biofuels change ownership (e.g., are sold by a producer to a blender), the RINs are also transferred. When a renewable fuel is blended or supplied for retail sale or at the port of embarkation for export, the RIN is separated from the fuel and maybe used for compliance or trade. The Code K status of the RIN is changed at separation.

Small Refinery Exemption

A permanent exemption is available to any parties who produce or import less than 10,000 gallons of renewable fuel in a year—they are not required to generate RINs for that volume, and are not required to register with the EPA if they do not take ownership of RINs generated by other parties. Under EISA, this exemption is temporarily extended (for up to three years beginning with the calendar year in which the refinery produces its first gallon of renewable fuel) to renewable fuel producers who produce less than 125,000 gallons per year from new production facilities. This exemption is intended to allow pilot and demonstration plants to focus on developing the technology and obtaining financing during their early stages rather than complying with RFS2 regulations.

Flexibility in Administering the RIN Requirements

RINs generated during the current year may be used to satisfy either the current year's or the following year's RVO. A RIN is not viable for any year's RVO beyond the immediately successive year; thus giving it essentially up to a two-year lifespan. For any individual company, up to 20% of the current year's RVO may be met by RINs from the previous calendar year.

In addition to compliance demonstration, RINs can be used for credit trading. When a fuel supplier has blended or sold a quantity of biofuel, the RINs are separated from the biofuels. If a supplier has already met its mandated share and has supplied surplus biofuels for a particular biofuel category, it can sell the extra RINs to another entity or it can hold onto the RINs for future use (either to satisfy the succeeding year's requirement or for sale in the succeeding year). An obligated party who faces a RIN deficit can purchase excess RINs. Since biofuels supply and demand can vary over time and across regions, a market has developed for RINs.

The marketability of RINs allows fuel suppliers who have not bought enough biofuels to fulfill their RFS requirement for each of the four RFS categories by purchasing the biofuels-specific RINs instead. As a result, RINs have value as a replacement for the actual purchase and supply of biofuels. Because four separate biofuel mandates must be met, the RIN value may vary across the individual biofuel categories.[43] Since the RFS biofuels categories are nested, the price of biodiesel RINs is generally equal to or greater than the price of RINs for advanced biofuels which, in turn is generally equal to or greater than the RIN value for total renewable biofuels. Thus, RIN values may vary across RFS categories as well as geographically with variations in specific biofuels supply and demand conditions.

Differences in RIN values also reflect the degree to which the mandate associated with a specific RIN biofuel category is binding on the market equilibrium.[44] For example, if the supply of a specific biofuel—including both domestically produced as well as imported—available to the market exceeds the RFS mandate (see left-hand side of **Figure 2** where $Q^* > Q_{RFS}$), then the RIN's "core" value (i.e., its price minus transaction costs and speculative component) would be zero at the mandated level (Q_{RFS}).

In contrast, if the mandated biofuel usage level exceeds what is offered by the market (see right-hand side of **Figure 2** where $Q_{RFS} > Q^*$), the biofuels mandate is binding because it forces obligated parties ("blenders" for this discussion)[45] to use more biofuels than they would without the mandate. The price of the biofuel purchased by the blender has to rise to $P_{producer}$ to solicit the extra production from the biofuels producers, while the biofuels price must fall to $P_{blender}$ to encourage greater blender purchases. The RIN's core value would be equal to the gap between these two prices, $P_{producer}$ minus $P_{blender}$. However, the blender must pay the full price of $P_{producer}$, which includes both $P_{blender}$ plus the RIN's core value, to acquire the mandated Q_{RFS}.

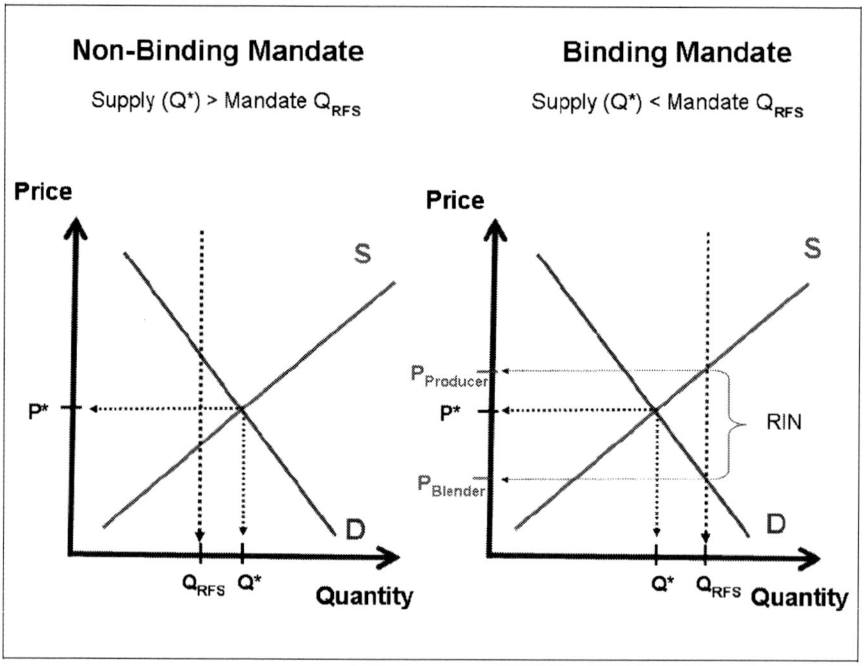

Source: "Renewable Identification Numbers are the Tracking Instrument and Bellwether of U.S. Biofuel Mandates," by Wyatt Thompson, Seth Meyer, and Pat Westhoff, *EuroChoices* 8(3), 2009.

Note: Supply equals domestic production and imports; demand equals both blenders and exporters demand.

Figure 2. How a Mandate May or May Not Affect RIN Values.

A RIN also may have speculative value, even when in surplus, if an investor were to anticipate a shortage in the near future (i.e., within the period for which a RIN is valid) and seek to acquire RINs cheaply in advance of the shortage. Prior to 2013, the overall biofuels mandates had not been binding and renewable fuel RIN values were small (usually less than $0.05/gallon). However, based on EIA data for U.S. gasoline consumption, it is expected that the RFS as scheduled in the statute will become binding sometime during 2013 or 2014 (see discussion later in this report). As a result, RIN prices escalated sharply in 2013 to prices well in excess of $1.00/gallon, before dropping dramatically in the second half of the year. Economists suggest that market conditions—supply, demand, and asymmetries in market power—will determine how any added cost of biofuels acquisition (i.e., the RIN value) would be distributed along the supply chain—that is, partially absorbed by biofuel producers and blenders or passed on to motor fuel consumers in the form of higher fuel prices.[46]

Because RINs have value, they are not immune to fraudulent activity. In late 2011 and early 2012, EPA issued notices of violations (NOVs) to three companies (Clean Green Fuels, LLC, Absolute Fuels, LLC, and Green Diesel, LLC) that the agency alleges fraudulently generated a combined 140 million biodiesel RINs in 2010 and 2011. Subsequently, individuals representing two of these companies have also faced criminal prosecution.[47]

EPA Analysis of RFS Impacts

As part of its final rule determination, EPA included an analysis of the market and environmental impact of the increased use of renewable fuels under the RFS2 standards.[48] The analytical results include the following.

- **Reduced dependence on foreign sources of crude oil.** By 2022, the mandated 36 bgals of renewable fuel will displace about 13.6 bgals of petroleum-based gasoline and diesel fuel, representing about 7% of expected annual U.S. transportation fuel consumption.
- **Reduced price of domestic transportation fuels.** By 2022, the increased use of renewable fuels is expected to decrease gasoline costs by $0.024 per gallon and diesel costs by $0.121 per gallon, producing a combined annual savings of nearly $12 billion.
- **Reduced GHG emissions.** When fully implemented in 2022, the expanded use of biofuels under the RFS is expected to reduce annual GHG emissions by 138 million metric tons—equivalent to taking about 27 million vehicles off the road.
- **Increased U.S. farm income.** By 2022, the expanded market for agricultural products such as corn and soybeans resulting from biofuels production is expected to increase annual net farm income by $13 billion.

- **Decreased corn and soybean exports.** The expanded use of corn starch and soybean oil for biofuels is expected to reduce corn exports by 8% and soybean exports by 14% by 2022.
- **Increased cost of food in the United States.** The increased demand for U.S. agricultural products is expected to raise the overall commodity price structure, leading to an annual increase in the cost of food per capita of about $10 by 2022, or over $3 billion.
- **Increased emissions of certain air contaminants, but decreased emissions of others.** Contaminants expected to increase include hydrocarbons, nitrogen oxides (NOx), acetaldehyde, and ethanol; those expected to decrease include carbon monoxide (CO) and benzene. The effects are expected to vary widely across regions, but in the net, increases in population-weighted annual average ambient PM and ozone concentrations are anticipated to lead to up to 245 cases of adult premature mortality.

RFS as Public Policy

Proponents' Viewpoint

Supporters of an RFS claim it serves several public policy interests[49] in that it:

- reduces the risk of investing in renewable biofuels by guaranteeing biofuels demand for a projected period (such risk would otherwise keep significant investment capital on the sidelines);
- enhances U.S. energy security via the production of liquid fuel from a renewable domestic source resulting in decreased reliance on imported fossil fuels (the United States currently imports over half of its petroleum,[50] two-thirds of which is consumed by the transportation sector);
- provides an additional source of demand—renewable biofuels—for U.S. agricultural output that has significant agricultural and rural economic benefits via increased farm and rural incomes and substantial rural employment opportunities;[51]
- underwrites the environmental benefits of renewable biofuels over fossil fuels (most biofuels are non-toxic, biodegradable, and produced from renewable feedstocks); and

- responds to climate change concerns because agricultural-based biofuels emit substantially lower volumes of direct greenhouse gases (GHGs) than fossil fuels when produced, harvested, and processed under certain circumstances.

Critics' Viewpoints

Critics of an RFS, particularly of the EISA expansion of the original RFS, have taken issue with many specific aspects of biofuels production and use, including the following:

- By picking the "winner," policymakers may exclude or retard the development of other, potentially preferable alternative energy sources.[52] Critics contend that biofuels are given an advantage via billions of dollars of annual subsidies that distort investment markets by redirecting venture capital and other investment dollars away from competing alternative energy sources. Instead, these critics have argued for a more "technology-neutral" policy such as a carbon tax, a capand-trade system of carbon credits, or a floor price on imported petroleum.
- Continued large direct and indirect federal incentives for corn-starch ethanol production are no longer necessary since the sector is no longer in its "economic infancy" and would have been profitable in most months since 2006 without federal subsidies.[53]
- The expanded mandate could have substantial unintended consequences in other areas of policy importance, including energy/petroleum security, pollutant and greenhouse gas emissions, agricultural commodity and food markets, land use patterns, soil and water quality, conservation, the ability of the gasoline-marketing infrastructure and auto fleet to accommodate higher ethanol concentrations in gasoline, the likelihood of modifications in engine design, and other considerations.
- Taxpayers are being asked to finance continued biofuels subsidies in support of existing and future biofuels infrastructure that have the potential to affect future federal budgetary choices.

Cost Estimates of Biofuels Policy

Historically, the major direct federal costs associated with the implementation of the RFS have been the federal tax credits available to the various biofuels that are blended to meet the RFS mandate.[54] Most of these tax credits expired at the end of 2011, while the cellulosic biofuels production tax credit is scheduled to expire at the end of 2013.

Prior to their expiration, the combination of biofuels tax credits with other federal and state government subsidies in support of ethanol production were estimated to be in the range of $5.4 billion to $6.6 billion per year—averaging nearly $1 per gallon of biofuel produced in the United States.[55] In 2011 (the last year in which the full suite of biofuels tax credits was in effect), federal subsidies were estimated at over $7.8 billion, including nearly $7.5 billion in tax credits.[56]

With the expiration of the ethanol tax credit, estimates of federal support have fallen sharply. In 2012, following the expiration of the corn ethanol tax credit, a preliminary estimate of federal biofuel subsidies (including Title IX farm bill energy programs)[57] was approximately $1.3 billion, of which slightly more than $1 billion was attributable to the biodiesel tax credit of $1.00 per gallon.

Both the cellulosic biofuels production tax credit of $1.01 per gallon and the biodiesel and renewable diesel fuel mixtures tax credit of $1.00 per gallon were extended through 2013 by the American Taxpayers Relief Act of 2012 (ATRA: P.L. 112-240). ATRA also retroactively applied the extension to biodiesel and renewable diesel fuel mixtures produced in 2012.

Proposed RFS-Related Legislation

There is currently no proposed legislation in the 113th Congress to renew or extend either of these biofuel tax credits. However, several bills propose to alter RFS mandates and EPA's waiver of E15 blending ratios.

Table 6. Selected Biofuels-Related Bills in the 113$_{th}$ Congress

Bill Number	Bill Name	Sponsor	Action
H.R. 550	Phantom Fuel Reform	Rep. Gregg Harper	To amend the RFS to require the cellulosic biofuel

Bill Number	Bill Name	Sponsor	Action
S. 251	Act of 2013	Sen. Flake	requirement to be based on actual production for the Jan.-Oct. period of the preceding year, pro-rated to an annual basis.
H.R. 596	Public Lands Renewable Energy Development Act of 2013	Rep. Paul Gosar	To promote the development of renewable energy on public lands.
H.R. 796	Amendment to the Clean Air Act	Rep. Sensenbrenner	To limit the cellulosic RFS mandate to be not more than 5% or 1 million gallons (whichever is greater) more than the total volume of cellulosic biofuel that was commercially available for the most recent calendar year.
H.R. 875	untitled	Rep. Sensenbrenner	To provide for a comprehensive assessment of the scientific and technical research on the implications of the use of mid-level ethanol blends (e.g., E15).
H.R. 979	Forest Products Fairness Act of 2013	Rep. Thompson	To modify the definition of the term "biobased product" to more broadly include forest products.
H.R. 1214	Domestic Fuels Protection Act of 2013	Rep. Shimkus	To provide liability protection for claims based on the design, manufacture, sale, offer for sale, introduction into commerce, or use of certain fuels and fuel additives (e.g., E15).

Table 6. (Continued)

Bill Number	Bill Name	Sponsor	Action
H.R. 1273	Rural Energy Improvement Act	Rep. Welch	To reauthorize and improve the Rural Energy for America Program (REAP).
H.R. 1461	RFS Elimination Act	Rep. Goodlatte	To repeal the RFS program of the EPA.
H.R. 1462	RFS Reform Act of	Rep. Goodlatte	To prohibit the EPA from approving the introduction
S. 344	2013	Sen. Wicker	into commerce of gasoline that contains greater than 10%-volume ethanol
H.R. 1469	Leave Ethanol Volumes at Existing Levels (LEVEL) Act	Rep. Burgess	To limit expansion of RFS biofuel mandates, to prohibit authorization of ethanol blends greater than 10%.
H.R. 1482	RFS Amendments Act	Rep. Womak	To eliminate corn ethanol requirements under the RFS program
S. 389	Freedom Fuels Act of 2013	Sen. Baucus	To authorize long-term contracts for the procurement of certain liquid transportation fuels for the Dept. of Defense

Source: Legislative Information System of the U.S. Congress.

Notes: This is not meant to serve as a comprehensive list of all energy-related bills, but instead represents a selection of bills deemed (by CRS) most relevant to federal biofuels programs and policies.

POTENTIAL ISSUES WITH THE EXPANDED RFS

Most U.S. biofuel production is ethanol produced from corn starch. As a result, as the U.S. ethanol industry has grown over the years, so too has its usage share of the annual corn crop. In 2000, national ethanol production was using about 6% of U.S. corn supplies; by 2012 it was expected to use about

31%.[58] The principal co-product from ethanol production—Distiller's Dried Grains or DDG—is useful as a relatively high-protein animal feed. About 30% (by weight) of corn used for biofuels is left over from the production process in the form of DDGs. The supply of DDGs has become more abundant in direct correlation with the expansion of the ethanol industry, and the U.S. livestock sector has learned how to incorporate this new feedstuff into animal and poultry rations, thus increasing both DDG value and average returns to ethanol production. Corn use for ethanol peaked at 5,019 million bushels in 2010, then declined sharply to 4,648 million bushels during the severe drought of 2012. USDA currently estimates 2013 corn use for ethanol at 4,900 million bushels.[59] Under the expanded RFS, the 2015 corn ethanol cap of 15 billion gallons, coupled with the existing U.S. ethanol production capacity of nearly 15 billion gallons, suggests that ethanol will likely use a declining share—perhaps in the 25% to 30% range—of the volume of U.S. corn production (adjusted for DDGs) in the future, depending on yield and area developments, and petroleum market conditions.

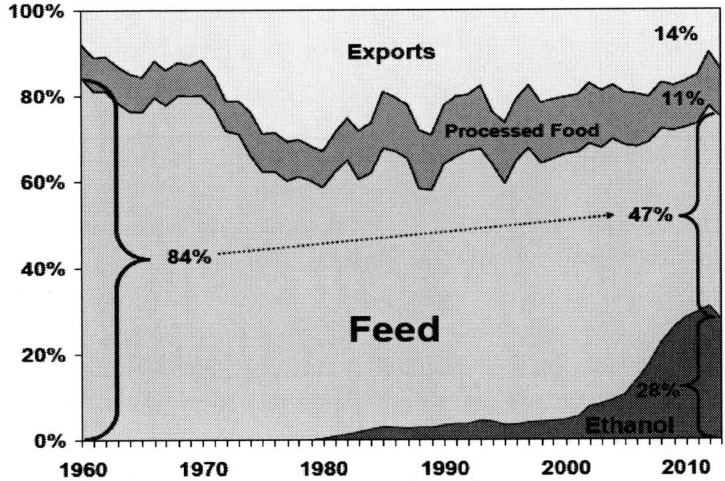

Source: USDA, Production, Supply, and Demand (PSD) database, November 8, 2013.
Note: Data are adjusted to account for use of ethanol by-products as animal feed; about 30% (by weight) of corn used for biofuels is left over from the production process and is useful as a relatively high-protein animal feed which is both used domestically and exported.

Figure 3. Ethanol Use Grew Rapidly as Share of U.S. Corn Supply from 2005 to 2012. (annual U.S. corn disappearance categories, as a % of total use, excluding ending stocks).

The shift towards greater corn use for biofuels that occurred from 2006 to 2010 meant higher prices for other corn users, including both the livestock and export sectors (**Figure 3**). The biofuels-driven expansion in feedstocks production (especially corn for grain and stover) has heightened competition for available cropland between biofuels feedstocks and other field crops, and has resulted in more intense agricultural activity on U.S. cropland to meet growing demand for food, feed, and fuel resources. This has consequences for several important agricultural markets, including

- grains—because corn competes with other grains for land;
- livestock—because animal feed costs increase with the price of corn;
- agricultural inputs—because corn is more input-intensive (in terms of fertilizers and pesticides) than other major field crops; and
- land—because the value of cropland, as well as total harvested acreage, increases with commodity prices and returns per acre.

As recently as 2004, over 60% of U.S. corn production was used as feed. The feed share of corn declined to 40% in 2011, but is expected to rebound slightly to 47% with the outlook for a record corn crop in 2013.[60]

Overview of Long-Run Corn Ethanol Supply Issues

The ability of the U.S. corn industry to continue to expand production and satisfy the steady growth in demand depends, first and foremost, on continued productivity gains. U.S. corn yields have shown strong, steady growth since the late 1940s, with some acceleration occurring since the mid-1990s as bio-engineered advances in seed technology have heightened drought and pest resistance in corn plants (**Figure 4**). Weather-related problems, including the severe drought of 2012, pulled corn yields well below trend; however, yields are estimated to return to trend in 2013. A return to normal growing conditions is expected to replenish corn supplies, lower prices, and return the supply and demand situation to a more traditional balance.

Corn Prices

From 1998 through early 2006, the U.S. farm price of corn averaged $2.09 per bushel. However, average corn prices jumped to $3.83 per bushel during the 2006-2010 period, when the rapid expansion of the U.S. ethanol industry exceeded the growth in U.S. corn production. Unfavorable weather reduced corn yields slightly below trend in both 2010 and 2011, but the severe drought of 2012 (**Figure 4**) dropped yields 23% below trend. The combination of these successive below-trend yield outcomes, in the face of continued strong demand for corn, pushed prices temporarily to a $6.29-per-bushel average during 2011 through early 2013 (**Figure 5**). The outlook for a return to trend yields and a record harvest of 14 billion bushels of corn in 2013 has since pressured prices back towards the $4.00 per bushel level. USDA projects corn prices to remain in the $4.00 to $5.00 per bushel range through 2020.[61]

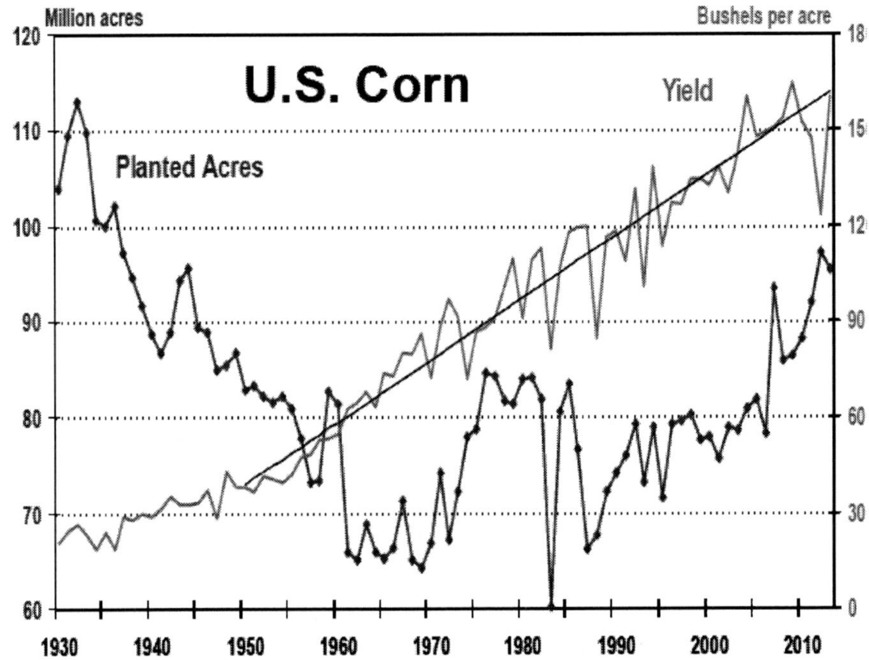

Source: USDA, PSD database, as of November 8, 2013. Yield trend line calculated by CRS.

Figure 4. U.S. Annual Corn Planted Acres and Yield.

Corn Yield and Area

It is likely that upward-trending farm prices (**Figure 5**) will encourage continued research investments to move corn yields steadily higher in the future. U.S. cropland planted to corn has increased in recent years from the 1983 low of 60.3 million acres to an estimated high of 97.2 million acres in 2012—the highest since 1936. Prospects for further expansion in crop area are far less certain, as corn is an energy-intensive crop that prefers deep, fertile soils and timely precipitation. Within the prime corn-growing regions of the Corn Belt, per-acre returns for corn easily dwarf other field crops that vie for the same acreage. Recent seed developments have allowed corn production to expand dramatically into the central and northern Plains states. However, the risk of investing up front in high operating costs to be offset at harvest by strong returns increases as production moves into less traditional regions, such as the northern Plains, the Delta, and the Southeast.

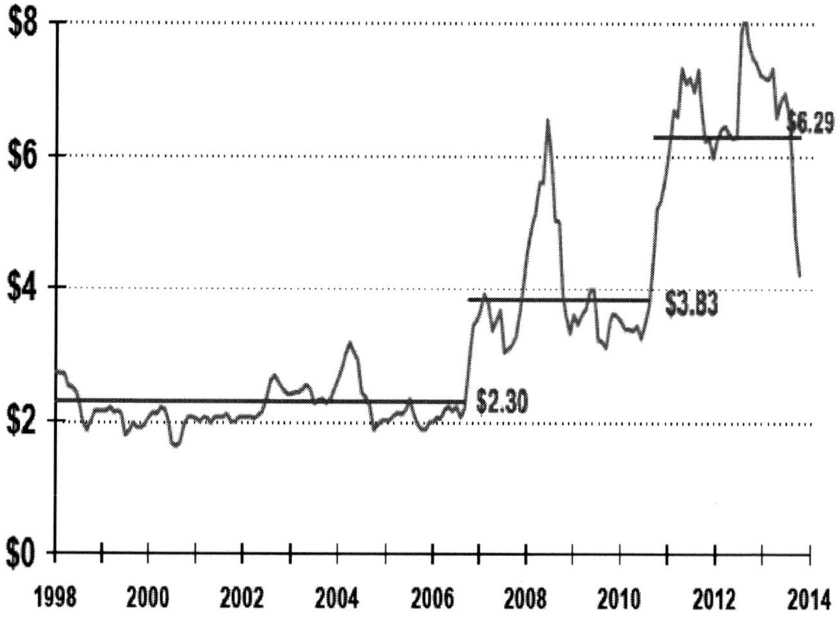

Source: USDA, ERS, Feed Grains Database, at http://www.ers.usda.gov/Data/feedgrains/; as of Nov. 22, 2013.

Figure 5. Monthly U.S. Corn Prices Have Trended Upward Since Late 2005. (central Illinois cash price for no. 2, yellow corn).

The most likely source of new corn acreage will come from shifts in crop rotation from soybeans to corn.[62] However, crop intensification also has its limits. Corn (of the grass family) is traditionally planted in an annual rotation with soybeans (a broad-leaf legume) that offers important agronomic benefits including pest and disease control, as well as enhanced soil fertility.[63] When farmers shift away from this rotation, corn yields tend to suffer. Planting successive corn crops—referred to as corn-on-corn cultivation—rather than in rotation with soybeans, wheat, or fallow generally produces a yield drag on successive corn crops that can lower yields anywhere from 5% to 15%, depending on soil, climate, and cultivation practices.[64] As a result, the corn-to-soybean price ratio would have to tilt fairly strongly in favor of corn for corn-on-corn production to be profitable.

Given the limitations on corn area expansion and rotational intensification, it is likely that the sustainable long-run corn planted area is probably in the range of 90 million to 95 million acres. If this is the case, then it would mean that future growth in U.S. corn production will be increasingly dependent on yield growth.

Overview of Non-Corn-Starch-Ethanol RFS Issues

EISA defines "advanced biofuels" very broadly as biofuels other than corn-starch ethanol that achieve a 50% reduction in greenhouse gas emissions relative to gasoline. As such, advanced biofuels would include imported Brazilian sugar-cane ethanol, as well as home-grown biodiesel. However, the principal focus of advanced biofuels is on biofuels based on cellulosic biomass. Under the RFS2, advanced biofuels use is mandated to reach a minimum of 21 billion gallons by 2022, of which at least 16 billion gallons must be some type of cellulosic biofuel. However, many obstacles must first be overcome before commercially competitive cellulosic biofuels production occurs.[65]

In the near term, it is likely that corn stover[66] will be a primary biomass of choice for cellulosic biofuels production. This is because many ethanol plants already exist in corn production zones and an extension of those plants to include cellulosic biofuels production from stover would offer some scale economies. However, stover-to-biofuel conversion has its own set of potential environmental trade-offs, paramount of which is the dilemma of sacrificing soil fertility gains by harvesting the stover rather than returning it to the soil under no- or minimum-tillage practices. Other potential near-term supplies include other agricultural and municipal wastes.

Cellulosic Biofuels Production Uncertainties

There are substantial uncertainties regarding both the costs of producing cellulosic feedstocks and the costs of producing biofuels from those feedstocks. Dedicated perennial crops are often slow to establish, and it can take several years before a marketable crop is produced. Crops heavy in cellulose tend to be bulky and represent significant problems in terms of harvesting, transporting, and storing. New harvesting machinery would need to be developed to guarantee an economic supply of cellulosic feedstocks.[67] Seasonality issues involving the operation of a biofuels plant year-round based on a four- or five-month harvest period of biomass suggest that storage is likely to matter a great deal. In addition, most marginal lands (i.e., the low-cost biomass production zones) are located far from major urban markets, making it difficult to reconcile plant location with the cost of fuel distribution.

Following the EPA's substantial revisions to the first four years of cellulosic biofuels mandates (2010-2013), there has been considerable uncertainty surrounding current cellulosic biofuel conversion technologies and the cost of the conversion process (including physical, chemical, enzymatic, and microbial treatment and conversion of the biomass feedstocks into motor fuel). These uncertainties, plus the financial crisis of 2008 and the ensuing recession and credit crunch, severely curtailed new investment in the biofuels sector.[68]

However, it appears that 2013 may experience the first substantial commercial production levels of cellulosic biofuels. EPA finalized a cellulosic biofuels RFS of 6 million gallons under the expectation that two plants would begin commercial production during 2013, with possibly one more plant expected to follow by the fourth quarter of 2013.[69] However, while production from at least one of the plants has started, through October 2013 fewer than 0.6 million cellulosic RINs were registered in EPA's system[70] – significantly more than in previous years, but still well below the 2013 mandate level.

However, cellulosic fuel supply may be poised to grow rapidly, at least in relative terms. Industry reports suggest that new cellulosic biofuels plants are either in the planning stages or under construction in as many as 20 states and Canadian provinces.[71]

Unintended Policy Outcomes of the ' ' Advanced Biofuels' ' Mandate

Because the advanced biofuels mandate in the RFS is a fixed mandate, irrespective of prices, the above uncertainties about the production of cellulosic ethanol could have significant implications for fuel supply and fuel prices. If cellulosic ethanol production is unable to advance rapidly enough to meet the RFS mandate for non-corn-starch ethanol, then other unexpected biofuels sources may be forced to step in and fill the void:

- production of domestic sorghum-starch ethanol may expand across the prairie states and in other regions less suitable for corn production;

- costly domestic sugar-beet ethanol or biodiesel production may be undertaken to fill the mandate; or
- imports of Brazilian sugar-cane ethanol could expand.

Energy Supply Issues

Biofuels are not primary energy sources. Energy is first stored in biological material (through photosynthesis), and then must be converted into a more useful, portable fuel. This conversion requires energy. The amount and types of energy used to produce biofuels (e.g., coal versus natural gas), and the feedstocks for biofuels production (e.g., corn versus cellulosic biomass), are critical in determining a biofuel's net energy balance and the environmental benefits of a biofuel.

Energy Balance

To analyze the net energy consumption of ethanol, the entire fuel cycle must be considered. The fuel cycle consists of all inputs and processes involved in the development, delivery, and final use of the fuel. For corn-based ethanol, these inputs include the energy needed to produce fertilizers, operate farm equipment, transport corn, convert corn to ethanol, and distribute the final product.

There are a wide range of estimates of the net energy output/input ratio for corn-starch ethanol production, although many of these are now dated. The most recent study by USDA estimated an energy output/input ratio of 2.3 based on a 2005 survey of corn growers and 2008 data for ethanol plants (and assuming the then-most-advanced technology for corn and ethanol production).[72] A 2.3 ratio implies that the energy contained in a gallon of corn ethanol was 130% higher than the amount of energy needed to produce and distribute it. Ethanol industry sources argue that technological innovation will continue to improve corn ethanol's energy balance. However, other analyses have resulted in a significantly lower output/input ratio.

If feedstocks other than corn are used to produce biofuels, it is expected that lower nitrogen fertilizer use would greatly improve the energy balance. Further, if biomass were used to provide process energy at the biofuels refinery (rather than coal or natural gas), the energy savings would be even greater.[73] Some estimates are that cellulosic ethanol could have an energy balance of 8.0 or more.[74] Similarly high energy balances have been calculated for sugar-cane ethanol and certain types of biodiesel.

Energy Security[75]

Ethanol displaces gasoline, and the benefits to energy security from ethanol, while relatively small, are still potentially important. However, biofuels' potential to play a larger role in energy security is questionable. Roughly 40% of the U.S. corn crop was used for ethanol in 2012, and the resultant ethanol accounted for about 7% of gasoline consumption on an energy-equivalent basis.[76] There is considerable uncertainty regarding how quickly or how much U.S. corn production can expand. If the entire 2012 U.S. corn crop of 10.8 billion bushels were used as ethanol feedstock, the resultant 30 billion gallons of ethanol (20.6 billion gasoline-equivalent gallons, or GEG) would represent about 16% of estimated national gasoline use of approximately 132 billion gallons.[77]

An expanded RFS would certainly displace petroleum consumption, but the overall effect on life-cycle fossil fuel consumption is questionable, especially if there is a large reliance on corn-based ethanol. Under the EISA RFS mandate, by 2022 biofuels will still represent about 20% of gasoline energy transportation fuel demand and 2.4% of diesel transportation fuel demand.[78]

The specific definition of "advanced biofuels" also affects the overall energy security picture for biofuels. For example, an expanded RFS provides an incentive to increase imports of sugar-cane ethanol, especially from Brazil. The expanded RFS may also provide an incentive for imports of biodiesel and other renewable diesel substitutes from tropical countries, although EPA has determined that biodiesel from palm oil does not meet the necessary greenhouse gas reductions. The supplies would represent a "diversification" of fuel sources, not the "domestication" that some claim is true energy security.

Energy Prices

The effects of the expanded RFS on energy prices are uncertain. If wholesale biofuels prices are higher than gasoline prices (after all economic incentives are taken into account), then mandating higher and higher levels of biofuels would likely lead to higher gasoline pump prices. However, if petroleum prices—and thus gasoline prices—are high, the use of some biofuels might help to mitigate high gasoline prices.

Current production costs are thought to be so high for some biofuels, especially cellulosic biofuels and biodiesel from algae, that significant technological advances—or significant increases in petroleum prices—would be necessary to make them competitive with gasoline.[79]

Without cost reductions, mandating large amounts of these fuels would likely raise fuel prices. If a price were placed on greenhouse gas emissions—perhaps through the enactment of a carbon tax—then the economics could shift in favor of these fuels despite their high production costs, as they have lower fuel-cycle and life-cycle greenhouse gas emissions (see below).

Ethanol Infrastructure and Distribution Issues

In addition to the above concerns about feedstock supply for ethanol production, there also are issues involving ethanol distribution and infrastructure. Expanding ethanol production likely will strain the existing supply infrastructure. Further, expansion of ethanol use beyond the current 10% blend will require investment in entirely new infrastructure that would be necessary to handle an increasing percentage of ethanol in gasoline, or retrofitting and recertification of existing equipment. On the other hand, if drop-in fuels (i.e., petroleum-like biofuels such as biobutanol or biomass-based diesel substitutes) are produced in large quantities, some of these infrastructure issues may be mitigated, since these fuels can be used in existing infrastructure.

Distribution Issues

Unlike petroleum products, ethanol and ethanol-blended gasoline cannot be shipped in existing U.S. pipeline infrastructure. Ethanol-blended gasoline tends to separate in pipelines due to the presence of water in the lines.[80] Further, ethanol is corrosive and may damage existing pipelines and storage tanks. Also, corn ethanol must be moved from rural areas in the Midwest to more populated areas, which are often located along the coasts. This shipment is in the opposite direction of existing pipeline transportation, which moves gasoline from refiners along the coasts to other coastal cities and into the interior of the country. While some studies have concluded that shipping ethanol or ethanol-blended gasoline via pipeline could be feasible, no major U.S. pipeline has made the investments to allow such shipments on a large scale.[81]

The current distribution system for ethanol is dependent on rail cars, tanker trucks, and barges. These deliver ethanol to fuel terminals where it is blended with gasoline before shipment via tanker truck to gasoline retailers. However, these transport modes lead to prices higher than for pipeline transport, and the supply of current shipping options (especially rail cars) is limited. Because of these distribution issues, some pipeline operators are seeking ways to make their systems compatible with ethanol or ethanol-blended gasoline. These modifications could include coating the interior of pipelines with epoxy or some other, corrosion-resistant material. Another potential strategy could be to replace all susceptible pipeline components with newer, hardier components. However, even if such modifications are technically possible, they likely will be expensive, and could further increase ethanol transportation costs.

As non-corn biofuels play a larger role, some of the supply infrastructure concerns may be alleviated. Cellulosic biofuels potentially can be produced from a variety of feedstocks that are more widely distributed throughout the country, unlike current dependency on a single crop (corn) from one region of the country. For example, municipal solid waste is ubiquitous across the United States, and could serve as a ready feedstock for biofuels production if the technology were developed to convert it economically to fuel. Further, increased imports of biofuels from other countries could allow for greater use of biofuels, especially along the coasts.

The Blend Wall and Higher-Level Ethanol Blends

It is now estimated that almost all gasoline sold in the United States contains some ethanol (mostly blended at the 10% level). A key benefit of gasoline-ethanol blends up to 10% ethanol is that they are compatible with existing vehicles and infrastructure (fuel tanks, retail pumps, etc.). All automakers that produce cars and light trucks for the U.S. market warrant their vehicles to run on gasoline with up to 10% ethanol (E10). As a result, this 10% blend has represented an upper bound (sometimes referred to as the "blend wall") to the amount of ethanol that can be introduced into the gasoline pool.[82] If most or all gasoline in the country contained 10% ethanol, this would allow only for roughly 13 billion gallons, far less than the RFS mandates for 2013 onward—13.8 billion gallons in 2013; 14.4 billion in 2014; and 15 billion in 2015.

For ethanol consumption to exceed the so-called "blend wall" and meet the RFS mandates, increased consumption at higher blending ratios is likely needed. For example, raising the blending limit from 10% to a higher ratio such as 15% or 20% would immediately expand the "blend wall" to somewhere in the range of 20 billion to 27 billion gallons. The U.S. ethanol industry is a strong proponent of raising the blending ratio.

In response to industry concerns regarding the impending blend wall, the EPA, after substantial vehicle testing, issued a partial waiver for gasoline that contains up to a 15% ethanol blend (E15) for use in model year 2001 or newer light-duty motor vehicles (i.e., passenger cars, light-duty trucks, and sport utility vehicles), but announced that no waiver would be granted for E15 use in model year 2000 and older light-duty motor vehicles, as well as in any motorcycles, heavy duty vehicles, or non-road engines.[83] According to the Renewable Fuel Association (RFA), the approval of E15 use in model year 2001 and newer passenger vehicles covered 62% of passenger vehicles on U.S. roads at the end of 2010.[84]

However, the EPA waiver for E15 is not sufficient, in and of itself, to ensure higher blending ratios. Fuel producers must also register new fuel blends and submit health effects testing to EPA. Further, numerous other changes have to occur before large numbers of gasoline stations will begin selling E15, including many approvals by states and potentially significant infrastructure changes (pumps, storage tanks, etc.). As a result, the vehicle limitation to newer models, coupled with infrastructure issues, is likely to limit rapid expansion of blending rates. EPA acknowledged this infrastructure limitation when it stated,[85]

> EPA recognizes that ethanol will likely continue to predominate the renewable fuel pool in the near future, and that for 2014 the ability of the market to consume ethanol in higher blends such as E85 is highly constrained as a result of infrastructure- and market-related factors. EPA does not currently foresee a scenario in which the market could consume enough ethanol sold in blends greater than E10, and/or produce sufficient volumes of non-ethanol biofuels to meet the volumes of total renewable fuel and advanced biofuel as required by statute for 2014.

Moreover, a group of engine and equipment manufacturers challenged the partial waiver in court, arguing that EPA failed to estimate the likelihood of misfueling (using E15 in equipment denied a waiver), and the economic and environmental consequences of that misfueling.[86] In response to these concerns, EPA requires E15 suppliers to submit to the agency misfueling mitigation plans (MMP).[87] Concerns over a preliminary MMP that required a four-gallon minimum purchase from some pumps supplying both E15 and E10 led to a new MMP that EPA approved in February 2013 that eliminates the four-gallon requirement as long as a fuel station has at least one dedicated E10 (or lower) pump to fuel older passenger cars and light trucks as well as non-road vehicles/engines.[88]

The blend wall problem is made more acute by substantial revisions in EIA's projections of U.S. transportation fuel consumption rates since the RFS was first passed into law in 2007 (**Figure 6**). At that time, EIA estimated that U.S. transportation consumers were using about 145 billion gallons of gasoline (including ethanol) per year, but that consumption would grow strongly to 176 billion gallons of gasoline by 2022—as a result, RFS mandated biofuels would represent about 19% of annual gasoline consumption. By 2013, EIA had substantially lowered its fuel consumption outlook—partly due to sustained high petroleum prices, the prolonged effects of the 2008 financial crises on consumer incomes, and significantly higher fuel economy standards on new vehicles. Instead of growth, EIA projects gasoline consumption to fall to about 120 billion gallons by 2022, thus causing the RFS mandate's share of the gasoline transportation fuel market (if left unchanged) to grow to nearly 20% of annual consumption (in gasoline-equivalent gallons).[89]

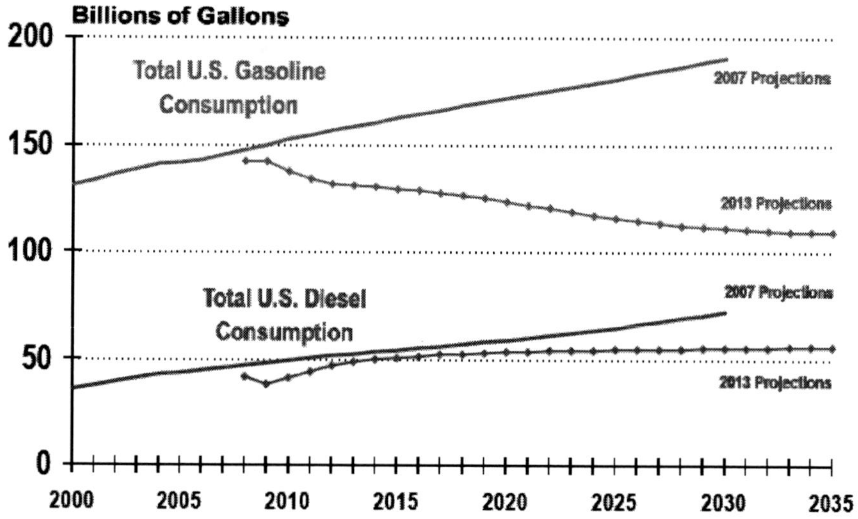

Source: DOE, EIA, Annual Energy Review 2007 and Annual Energy Review 2013.

Figure 6. EIA Long-Term Projections of U.S. National Transportation Fuel Use.

Two additional options to resolving this bottleneck exist but appear to be long-run alternatives. First, increased use of ethanol in flex-fuel vehicles (FFVs) at ethanol-to-gasoline blend ratios as high as 85% (referred to as E85) is a possibility. However, increased E85 use involves substantial infrastructure development, particularly in the number of designated storage tanks and E85 retail pumps, as well as a rapid expansion of the FFV fleet to absorb larger volumes of ethanol. Infrastructure expansion will require significant investments, especially at the retail level. Installation of a new E85 pump and underground tank can cost as much as $100,000 to $200,000.[90] However, if existing equipment can be used with little modification, the cost could be less than $10,000.

A second alternative is to expand use of processing technologies at the biofuel plant to produce biofuels in a "drop-in" form (e.g., butanol) that can be used by existing petroleum-based distribution and storage infrastructure and the current fleet of U.S. vehicles. However, more infrastructure-friendly biofuels generally require more processing than ethanol and are therefore more expensive to produce.

Vehicle Infrastructure Issues

As was stated above, if a large portion of any increased RFS is met using ethanol, then the United States likely does not have the vehicles to consume the fuel. The 10% blend wall on ethanol in gasoline for conventional vehicles still poses a significant barrier to expanding ethanol consumption beyond 14 billion gallons per year.[91] To allow more ethanol use, vehicles will need to be certified and warranted for higher-level ethanol blends, or the number of ethanol FFVs will need to increase. Turnover of the U.S. automobile fleet has slowed during the recession, making it more difficult to integrate FFVs into the fleet.

Ethanol RINs and the Blend Wall

In the face of a looming blend wall, the status of available RIN stocks of renewable and advanced biofuels for use by obligated parties towards mandate compliance remains an important issue.[92] Current RIN stocks in mid-2013 are estimated at approximately 2.5 billion gallons, of which 2 billion gallons are D6 RINs which can only be used for compliance with the renewable portion of the RFS mandate. However, RIN stocks were expected to tighten substantially in 2014 as a large portion of existing RINs is used for current mandate compliance as well as a scheduled increase in the RFS mandates for all categories in 2014. As a result, the price of renewable fuel RINs increased dramatically in the first half of 2013. Spot prices for ethanol RINs averaged between $0.07 and $0.08 per gallon in the first weeks of January. However, in the first week of March 2013, ethanol RINs averaged roughly $0.76 per gallon—a nine-fold increase.[93] By July 2013, RIN prices have pushed past $1.00 per gallon. In the second half of the year RIN prices dropped steadily with the expectation that EPA might use its waiver authority to lower the 2014 mandates below the 2013 levels, as the agency ultimately proposed in November 2013.

Conclusion

There is continuing interest in expanding the U.S. biofuels industry as a strategy for promoting energy security and achieving environmental goals. However, it is possible that increased biofuel production may place desired policy objectives in conflict with one another. There are limits to the amount of biofuels that can be produced from current feedstocks, particularly corn, and questions about the net energy and environmental benefits they might provide. Further, rapid expansion of biofuels production based on traditional field crops such as corn may have many unintended and undesirable consequences for agricultural commodity costs, fossil energy use, and environmental degradation. Owing to these concerns, alternative strategies for energy conservation and alternative energy production (including biofuels from nontraditional feedstock sources) are widely seen as warranting consideration.

End Notes

[1] For more information, see CRS Report R41282, Agriculture-Based Biofuels: Overview and Emerging Issues and CRS Report R41985, Renewable Energy Programs and the Farm Bill: Status and Issues.

[2] For more information on incentives (both tax and non-tax) for biofuels, see CRS Report R42566, Alternative Fuel and Advanced Vehicle Technology Incentives: A Summary of Federal Programs.

[3] In gasoline-equivalent shares with 7.3% for ethanol and 2.1% for biodiesel. CRS estimates based on data from Energy Information Agency (EIA), Department of Energy (DOE).

[4] Heating oil, jet fuel, and fuels for ocean-going vessels are excluded from RFS2's national transportation fuel supply; however, renewable fuels used for these purposes may count towards the RFS2 mandates. EPA, 40 C.F.R. Part 80, "Regulation of Fuels and Fuel Additives: Changes to Renewable Fuel Standard Program, Final Rule," Feb. 3, 2010.

[5] Ibid.

[6] CRS Report R40460, Calculation of Lifecycle Greenhouse Gas Emissions for the Renewable Fuel Standard (RFS).

[7] CRS Report R40529, Biomass: Comparison of Definitions in Legislation Through the 112th Congress.

[8] For more information, see the section "Implementation of the RFS" later in this report.

[9] "Renewable Fuels: Regulations & Standards," EPA's online chronicle of RFS rule making, available at http://www.epa.gov/otaq/renewablefuels/regulations.htm.

[10] The term "advanced biofuels" comes from legislation in the 110th Congress, and is defined in Section 201 of the Energy Independence and Security Act of 2007 (EISA). EISA defines "advanced biofuels" as biofuels other than ethanol derived from corn starch (kernels) having 50% lower lifecycle greenhouse gas emissions relative to gasoline. In some cases,

the definition of "advanced biofuels" includes mature technologies and fuels that are currently produced in large amounts. For example, the EISA definition of "advanced biofuels" includes ethanol from sugar cane, despite the fact that Brazilian sugar growers have been producing fuel ethanol for decades.

[11] As part of its February 3, 2010, final rule, EPA announced a revision in the BBD standard for 2010 to 1.15 bgals. This revision represents a summation of the 2009 standard of 0.5 bgals with the 2010 standard of 0.65 bgals. The RFS1 regulatory system, which was in effect during 2009 and which was based on national gasoline supply, did not provide any mechanism for implementing the 2009 BBD standard. As a result, it was integrated into the 2010 standard. Qualifying RINs accumulated during 2009 were acceptable in compliance.

[12] A diesel fuel product produced from cellulosic feedstocks that meets the 60% GHG threshold can qualify as either cellulosic biofuel or BBD.

[13] In addition, certain advanced biofuels have ethanol equivalency values greater than one. See the note on equivalency values later in this report. As a result, there is additional incentive to produce higher nested biofuels in meeting advanced and total biofuels RFS mandates.

[14] CRS Report R40460, Calculation of Lifecycle Greenhouse Gas Emissions for the Renewable Fuel Standard (RFS).

[15] EISA (P.L. 110-140), Title II, §201 Definitions, "(H) Lifecycle Greenhouse Gas Emissions."

[16] For more information on EPA's determination of lifecycle GHG emissions see CRS Report R40460, Calculation of Lifecycle Greenhouse Gas Emissions for the Renewable Fuel Standard (RFS).

[17] See "Section V. Lifecycle Analysis of Greenhouse Gas Emissions," Preamble, EPA RFS2 Final Rule, February 3, 2010, at http://epa.gov/otaq/renewablefuels/rfs2-preamble.pdf.

[18] For information on adding a potential new biofuel feedstock or production process pathway, see EPA, "Guidance on New Fuel Pathway Approval Process" at http://www.epa.gov/otaq/fuels/renewablefuels/compliancehelp/rfs2-lcapathways.htm.

[19] EPA, "EPA Issues Final Rule Additional Qualifying Renewable Fuel Pathways Under the Final RFS2 Program," EPA-420-F-13-014, February 2013.

[20] CRS Report R40529, Biomass: Comparison of Definitions in Legislation Through the 112th Congress.

[21] From various EPA announcements on "Renewable Fuels Regulations and Standards," at http://www.epa.gov/otaq/ fuels/renewablefuels/regulations.htm.

[22] EPA, 40 C.F.R. Part 80, "Regulation of Fuels and Fuel Additives: Changes to Renewable Fuel Standard Program, Final Rule," Feb. 3, 2010. EPA's official "Renewable Fuel Standard (RFS)" website, with links to all official documents, is available at http://www.epa.gov/otaq/fuels/renewablefuels/.

[23] The data are taken from EIA's October issue of its monthly Short-Term Energy Outlook Report, "Table 4a. U.S. Crude Oil and Liquid Fuels Supply, Consumption, and Inventories," and "Table 8. U.S. Renewable Energy Consumption," available at http://www.eia.gov/forecasts/steo/.

[24] EPA, 40 C.F.R. Part 80, Final Rule, Feb. 3, 2010, p. 14675. However, EPA was unable to meet the required timetable in announcing the 2013 Renewable Fuel Standards. Instead of November 2012, the final 2013 RFS was announced on August 6, 2013. Further, the proposed rule for 2014 was signed by EPA Administrator Gina McCarthy on November 15, 2013. The comment period runs for 60 days after the proposed rule is published in the Federal Register. However, the proposal had not been published as of November 22, 2013, so a final rule for 2014 is not expected until at least late January 2014. EPA Fact Sheet, "EPA Finalizes 2013 Renewable Fuel Standards," EPA-420- F-13-042, August 2013; and

"EPA Proposes 2014 Renewable Fuel Standards, 2015 Biomass-Based Diesel Volumes," EPA-420-F-13-048, November 2013.

[25] U.S. Code, "Regulation of fuels," 42 U.S.C. 7545.

[26] However, EPA has missed this statutory deadline in multiple years. For 2013, EPA did not even propose a standard until February 2013, with a public comment period running through late March.

[27] For more information, see CRS Report RS22870, Waiver Authority Under the Renewable Fuel Standard (RFS).

[28] New Mexico, Georgia, Texas, Virginia, Maryland, Delaware, Utah, and Wyoming. The governor of Florida wrote EPA in support of the waiver requests. 77 Federal Register 70754.

[29] See, "EPA Decision on Texas Request for Waiver of Portion of Renewable Fuel Standard (RFS)," EPA 420-F-08- 029, August 2008; at http://www.epa.gov/otaq /renewable fuels/420f08029.htm.

[30] The 2010 RFS was revised as part of a final rulemaking implementing the RFS as expanded by EISA, available at http://www.epa.gov/otaq/renewablefuels/420f10007.pdf. It should be noted that under the RFS1, fuels produced from feedstocks other than cellulose, but produced at facilities that used biomass for process energy, were treated as "cellulosic biofuels" for compliance purposes. These fuels were excluded from the definition of cellulosic biofuels in the EISA amendments to the RFS.

[31] EPA finalized the 2011 requirements in November 2010. EPA, "Regulation of Fuels and Fuel Additives: 2011 Renewable Fuel Standards; Final Rule," 75 Federal Register 76790-76830, Dec. 9, 2010.

[32] EPA finalized the 2012 requirements in December 2011. EPA, "Regulation of Fuels and Fuel Additives: 2012 Renewable Fuel Standards; Final Rule," 77 Federal Register 1320-1358, Jan. 9, 2012. EPA has proposed further revising this requirement to zero in response to the decision in API v. EPA vacating the 2012 standard (next page).

[33] EPA finalized the 2013 requirements in August 2013. EPA, "EPA Finalizes 2013 Renewable Fuel Standards," EPA-420-F-13-042, August 2013.

[34] This is the number of cellulosic RINs generated under the RFS2 program in 2012 as reported by EPA (as of March 1, 2013) on its "RFS2 EMTS Informational Data" online reporting system at http://www.epa.gov/otaq/fuels/rfsdata/.

[35] See "The Blend Wall and Higher-Level Ethanol Blends."

[36] EPA, 2014 Standards for the Renewable Fuel Standard Program; Proposed Rule (Prepublication Version), EPAHQ-OAR-2013-0479, Washington, DC, November 15, 2013, p. 90, http://www.epa.gov/otaq/fuels/renewablefuels/ documents/420f13048.pdf.

[37] 3.75 billion gallons – 2.20 billion gallons = 1.55 billion gallons.

[38] 18.15 billion gallons – 15.21 billion gallons = 2.94 billion gallons.

[39] "EPA Authority to Reduce the RFS", farmdocdaily.com, November 6, 2013, available at http://farmdocdaily.illinois.edu/pdf/fdd061113.pdf.

[40] EPA, "Update—2012 Cellulosic Biofuel Standard Mandate Issued," EnviroFlash, February 27, 2013. As part of the news release, EPA announced that since the 2012 mandate was zero, no compliance was necessary and any parties who had already submitted payment for 2012 cellulosic biofuel waiver credits would be issued refunds.

[41] EPA, 2014 Standards for the Renewable Fuel Standard Program; Proposed Rule (Prepublication Version), EPAHQ-OAR-2013-0479, Washington, DC, November 15, 2013, p. 24, http://www.epa.gov/otaq/fuels/renewablefuels/ documents/420f13048.pdf.

[42] See CRS Report R42824, Analysis of Renewable Identification Numbers (RINs) in the Renewable Fuel Standard (RFS). Other sources include Robert Wisner, "Renewable Identification Numbers (RINs) and Government Biofuels Blending Mandates," AgMRC Renewable Energy Newsletter, Agricultural Marketing Research Center, Iowa State University, April 2009; or Wyatt Thompson, Seth Meyer, and Pat Westhoff, "Renewable Identification Numbers are the Tracking Instrument and Bellwether of U.S. Biofuel Mandates," EuroChoices 8(3), 2009, pp. 43-50.

[43] For example, this was the case in late 2010. The biodiesel production tax credit of $1.00 per gallon had expired at the end of 2009 and subsequent biodiesel production dropped in 2010, potentially below that needed to meet the combined 2009/2010 mandate. As a result, biomass-based diesel RINs were trading at dramatically higher levels than one year previously. By the second week in December 2010, biomass-based diesel RINs were trading at $0.85 to $0.90 as compared to $0.01 to $0.02 the same week in 2009. However, with the enactment of an extension of the tax credit, biomass-based diesel RIN prices dropped by more than 30% in one week. "RIN Quotes," The Ethanol Monitor, vol. 6, no. 48 (December 20, 2010), p. 12.

[44] This discussion is based on "Renewable Identification Numbers are the Tracking Instrument and Bellwether of U.S. Biofuel Mandates," by Wyatt Thompson, Seth Meyer, and Pat Westhoff, EuroChoices 8(3), 2009.

[45] There is no explicit requirement in the RFS statute or regulations to blend biofuels into conventional fuels. However, the vast majority of compliance with the RFS mandates is the use of ethanol blended into motor gasoline.

[46] For a series of articles discussing the RFS mandate, RIN prices, and the blend wall, readers are referred to the policy articles authored by Nick Paulson, Scott Irwin, and Darrel Good (Dept. of Agriculture and Consumer Economics, Univ. of Illinois Urbana-Champaign) at http://farmdocdaily.illinois.edu/areas/policy/; Bruce Babcock and colleagues (Center for Agricultural and Rural Development, Iowa State Univ.) at http://www.card.iastate.edu/; or Pat Westhoff, Wyatt Thompson, and colleagues at the Food and Agricultural Policy Research Institute, Univ. of Missouri at http://www.fapri.missouri.edu/.

[47] For more information, see CRS Report R42824, Analysis of Renewable Identification Numbers (RINs) in the Renewable Fuel Standard (RFS).

[48] EPA, "Renewable Fuel Standard Program (RFS2) Regulatory Impact Analysis," Assessment and Standards Division, Office of Transportation and Air Quality, EPA-420-R-10-006, February 2010.

[49] See Renewable Fuels Assoc. (RFA), Position Papers at http://www.ethanolrfa.org/pages /position-papers.

[50] DOE, EIA, Annual Energy Review 2013, Table A1, "Total Energy Supply and Disposition Summary," Washington, December 5, 2012, at http://www.eia.gov/forecasts/aeo/er /pdf/tbla1.pdf.

[51] For example, see John M. Urbanchuk (Technical Director, Environmental Economics), Contribution of the Ethanol Industry to the Economy of the United States, white paper prepared for Renewable Fuels Assoc., January 31, 2013.

[52] For example, see Bruce A. Babcock, "High Crop Prices, Ethanol Mandates, and the Public Good: Do They Coexist?" Iowa Ag Review, Vol. 13, No. 2, Spring 2007; and Robert Hahn and Caroline Cecot, "The Benefits and Costs of Ethanol," Working Paper 07-17, AEI-Brookings Joint Center for Regulatory Studies, November 2007.

[53] AgMRC (agricultural marketing resource cneter), "Renewable Energy & Climate Change," Ethanol Profitability Spreadsheet, Iowa State University Extension and Outreach, November 6, 2013.

[54] See CRS Report R42566, Alternative Fuel and Advanced Vehicle Technology Incentives: A Summary of Federal Programs.

[55] Based on a 2007 survey of federal and state biofuel subsidies as reported in Ronald Steenblik, Biofuels—At What Cost? Government Support for Ethanol and Biodiesel in the United States, Global Subsidies Initiative of the International Institute for Sustainable Development, Geneva, Switzerland, September 2007, p. 37; available at http://www.globalsubsidies.org.

[56] CRS projection based on available data various USDA and DOE sources.

[57] See CRS Report R41985, Renewable Energy Programs and the Farm Bill: Status and Issues.

[58] The percentage is 42% prior to adjusting for the share of Distiller's Dried Grains or DDG contained in the corn used for ethanol that is diverted to animal feed. World Agricultural Outlook Board (WAOB), USDA, World Agr. Supply and Demand Estimates (WASDE), Nov. 8, 2013.

[59] Ibid.

[60] WAOB, USDA, WASDE, Aug. 12, 2013.

[61] USDA Agricultural Projections to 2022, OCE-2013-1, Office of the Chief Economist, February 2013.

[62] Chad E. Hart, "Feeding the Ethanol Boom: Where Will the Corn Come From?" Iowa Ag Review, vol. 12, no. 4 (Fall 2006), pp. 4-5.

[63] Bruce A. Babcock and David A. Hennessy, "Getting More Corn Acres from the Corn Belt," Iowa Ag Review, vol. 12, no. 4 (Fall 2006), pp. 6-7.

[64] Michael Duffy and David Correll, "The Economics of Corn on Corn," Integrated Crop Management, IC-498 (1), February 12, 2007.

[65] See CRS Report R41106, Meeting the Renewable Fuel Standard (RFS) Mandate for Cellulosic Biofuels: Questions and Answers, and CRS Report R41460, Cellulosic Ethanol: Feedstocks, Conversion Technologies, Economics, and Policy Options.

[66] Stover is the above-soil part of the corn plant excluding the kernels.

[67] To economically supply field residues to biofuels producers, farm equipment manufacturers likely would need to develop one-pass harvesters that could collect and separate crops and crop residues at the same time.

[68] CRS Report R41106, Meeting the Renewable Fuel Standard (RFS) Mandate for Cellulosic Biofuels: Questions and Answers.

[69] EPA, "Regulation of Fuels and Fuel Additives: 2013 Renewable Fuel Standards; Final Rule," August 6, 2013, at http://www.epa.gov/otaq/fuels/renewablefuels/regulations.htm.

[70] EPA, 2013 RFS2 Data, http://www.epa.gov/otaq/fuels/rfsdata/2013emts.htm.

[71] For example see, Advanced Ethanol Council (AEC), "Cellulosic Plants: Industry Progress Report 2012/13," December 19, 2012; and "Visible Progress in Commercialization of Advanced Biofuels, Biobased Products and Renewable Chemicals," Biotechnology Industry Organization (BIO), June 15, 2012.

[72] H. Shapouri, Paul W. Gallagher, Ward Nefstead, Rosalie Schwartz, Stacey Noe, and Roger Conway, 2008 Energy Balance for the Corn-Ethanol Industry, AER No. 846, Office of the Chief Economist, USDA, June 2010; hereinafter referred to as Shapouri et al. (2010).

[73] "Ethanol Energy Balance," Alternative Fuels & Advanced Vehicles Data Center, Dept. of Energy, available at http://www.afdc.energy.gov/afdc/ethanol/balance.html.

[74] David Andress, Ethanol Energy Balances, November 2002.

[75] A key question in evaluating the energy security benefits or costs of an expanded RFS is "what is the definition of energy security." For many policymakers, "energy security" and "energy independence" (i.e., producing all energy within our borders) are synonymous. For others, "energy security" means guaranteeing that we have reliable supplies of energy regardless of their origin. For this section, the former definition is used.

[76] By volume, ethanol accounted for nearly 10% of gasoline consumption in the United States in 2012, but a gallon of ethanol yields only about 68% of the energy of a gallon of gasoline.

[77] This estimate is based on DOE, EIA, Annual Energy Review 2013, Table A1, "Total Energy Supply and Disposition Summary," Washington, December 5, 2012,, and USDA's February 8, 2013, WASDE Report, using comparable conversion rates.

[78] Calculated by CRS based on EIA's 2013 Energy Outlook and the assumption that the BBD RFS remains fixed at 1.28 bgals through 2022.

[79] Plant-level cost-of-production (COP) data is proprietary and not readily available to the general public. News reports vary widely in their estimates of per-gallon COP for cellulosic biofuels. Furthermore, since commercial production of cellulosic biofuels is as yet minimal, most COP estimates are based on laboratory data which may not be representative of commercial scale costs.

[80] John Whims, "Pipeline Considerations for Ethanol," AgMRC, Sparks Companies, Inc., August 2002.

[81] For a DOE study of the economic feasibility of a hypothetical ethanol pipeline linking large East Coast demand centers with a stable supply from the Midwest, see "Report to Congress: Dedicated Ethanol Pipeline Feasibility Study," March 2010, at http://www1.eere.energy.gov/biomass/pdfs/report_to_congress_ethanol_pipeline.pdf.

[82] See CRS Report R40445, Intermediate-Level Blends of Ethanol in Gasoline, and the Ethanol "Blend Wall."

[83] For details, see EPA online information site, "E15 (a blend of gasoline and ethanol)," at http://www.epa.gov/otaq/ regs/fuels/additive/e15/.

[84] "E15 Decision Opens Blend to 2 Out of 3 Vehicles; More Work Yet to be Done," RFA news release, Jan. 21, 2011.

[85] EPA, "EPA Finalizes 2013 Renewable Fuel Standards," EPA-420-F-13-042, August 2013.

[86] Outdoor Power Equipment Institute (OPEI), Fact Sheet: E-15 Partial Waiver Legal Challenge, December 17, 2010. The case is Alliance of Automobile Manufacturers et. al v. Environmental Protection Agency. In August 2012 the D.C. Circuit held that the petitioners lacked standing because no engine or equipment owners had been injured in fact because little or no E15 had been introduced into commerce.

[87] See EPA online information site, "E15: Misfueling Mitigation Plans," at http://www.epa.gov /otaq/regs/fuels/ additive/e15/e15-mmp.htm.

[88] Ibid.

[89] Data are from EIA/DOE's 2013 Annual Energy Outlook. EIA also projects the U.S. national biodiesel transportation fuel market to show slow but steady growth (at about 1% per year) from about 47 bgals in 2012 to nearly 54 bgals by 2022. As a result, RFS BBD's share of the biodiesel transportation fuel market is projected to remain steady at about 2.5% through 2022.

[90] David Sedgwick, Automotive News, January 29, 2007. p. 112.

[91] Note that 15 billion gallons is the corn starch ethanol limit for the expanded RFS in the EISA.

[92] Nick Paulson, "RIN Update: Advanced and Renewable RIN Generation in 2013," http://farmdocdaily.com, July 31, 2013.

[93] "Ethanol & Gasoline Component Spot Market Prices," OPIS Ethanol and Biodiesel Information Service, various dates.

In: The Renewable Fuel Standard
Editor: Lydia Rider
ISBN: 978-1-63321-829-1

Chapter 2

WAIVER AUTHORITY UNDER THE RENEWABLE FUEL STANDARD (RFS)*

Brent D. Yacobucci

SUMMARY

Transportation fuels are required by federal law to contain a minimum amount of renewable fuel each year. The renewable fuel standard (RFS), established by the Energy Policy Act of 2005 (EPAct, P.L. 109-58) and amended by the Energy Independence and Security Act of 2007 (EISA, P.L. 110-140), required that a total of 16.55 billion gallons of renewable fuels be used to offset gasoline and diesel fuel in 2013. Under EISA, the scheduled mandates grow each year (to 36 billion gallons in 2022). However, the ability of fuel suppliers to meet the growing RFS mandates has been questioned. For 2014, instead of a total of 18.15 billion gallons scheduled in EISA, the Environmental Protection Agency (EPA) has proposed using waiver authority within the statute to lower the mandate to 15.21 billion gallons. Within the overall RFS there are sub-mandates (or "carveouts") for advanced and cellulosic biofuels. EPA has proposed lowering these as well.

Most of the overall RFS mandate (84% for 2013) is currently met using corn-based ethanol. The vast majority of the ethanol supplied to meet the RFS is blended into gasoline at the 10% level (E10). However,

---

* This is an edited, reformatted and augmented version of a Congressional Research Service publication RS22870, prepared for Members and Committees of Congress, dated January 21, 2014.

there is a limit to the amount of ethanol that can be blended into gasoline at this level, and other options for supplying ethanol (e.g., higher-level ethanol blends such as E15 and E85) are currently constrained because of infrastructure and other impediments. In prior years this had not been a problem because the volume of biofuels needed to meet the RFS mandate was below the maximum amount of ethanol that could be supplied as E10; but stakeholders were especially concerned about the possibility of facing the "blend wall" in either 2013 or 2014.

The Environmental Protection Agency (EPA) has the authority to waive the RFS requirements, in whole or in part, if certain conditions outlined in the law are present. Under EISA the overall RFS is scheduled to grow to 18.15 billion gallons in 2014. In November 2013 EPA proposed using its general authority to lower the RFS mandates in cases of "inadequate supply." In this case, they interpreted "inadequate supply" to include an inability to *deliver* the necessary biofuels to consumers, even if there is sufficient *production* of said biofuels. Gasoline and diesel fuel suppliers and other stakeholders have generally been supportive of EPA's proposal, while biofuel producers and corn growers believe that EPA is overstepping its authority.

Questions have also been raised over whether the overall mandate diverts enough corn supply from food/feed production to dramatically raise prices in those markets, and whether there is enough feedstock supply and production capacity to meet the carveouts for fuels other than corn ethanol. In 2008 the governor of Texas requested a waiver of the RFS because of high grain prices, although that waiver request was denied because EPA determined that the RFS requirements alone did not "severely harm the economy of a State, a region, or the United States," a standard required by the statute. A similar waiver petition was filed by the governors of several states in August 2012. That petition was denied for similar reasons.

In February 2010, as part of a final rulemaking implementing the RFS as expanded by EISA, EPA waived most of the 2010 cellulosic biofuel carveout—as they have in each subsequent year. In cases where EPA projects inadequate production of cellulosic biofuels to meet the schedule in EISA, the agency must lower the mandate to the projected level, although EPA's process for doing so has been controversial, and the agency's 2012 cellulosic mandate was remanded by the U.S. Court of Appeals for the D.C. Circuit.

This report provides a brief overview of the RFS program and discusses the process and criteria for EPA to waive various parts of the RFS.

BACKGROUND AND RECENT HISTORY

The Energy Policy Act of 2005 (EPAct, P.L. 109-58) established a renewable fuel standard (RFS), requiring the use of biofuels (such as ethanol) in the nation's fuel supply.[1] The Energy Independence and Security Act of 2007 (EISA, P.L. 110-140) significantly expanded this mandate. The RFS mandate has been a major impetus to the development of U.S. biofuels industries, especially the ethanol industry. As a result, ethanol production capacity and the demand for corn as a biofuel feedstock have grown dramatically over the past few years. In 2005, the United States produced 3.9 billion gallons of ethanol, requiring roughly 1.4 billion bushels of corn; in 2007, those numbers had increased to 6.5 billion gallons and 2.3 billion bushels. In 2007, roughly one-quarter of the U.S. corn crop was directed to ethanol production. In 2011, production had increased to roughly 14 billion gallons using approximately 5 billion bushels of corn or 40% of the 2011 crop.

Increasing demand for corn for biofuels, the rise in energy prices, and other supply concerns in international grain markets led to rapid increases in corn and other grain prices in 2008. These higher grain prices raised concern globally over food prices and availability. Because of these concerns, there was interest among some policymakers to amend or eliminate the RFS.

Under the provisions of EPAct and EISA, the administrator of the Environmental Protection Agency (EPA) has the authority to waive the RFS requirements in whole or in part, in response to a petition by a state or a fuel provider, or on her own motion. On April 25, 2008, Texas Governor Rick Perry sent a letter to EPA Administrator Stephen Johnson petitioning for a 50% waiver from the RFS requirements.[2] In his letter, Governor Perry stated that he initiated the petition because of the negative effect of the requirements on the Texas economy and on global food prices. In August 2008 EPA denied the waiver request because the agency found that the effects of the RFS on food, feed, and fuel prices was minimal, and thus the economic effects of the RFS "could not be categorized as severe."[3] In August 2012, Arkansas governor Mike Beebe and North Carolina governor Beverly Eaves Perdue submitted petition letters to EPA requesting full or partial waivers of the 2012 and 2013 RFS mandates.[4] Governors of several other states joined that petition.[5] The governors cited the effects of the 2012 drought, further compounded by the increased grain demand from the RFS, as leading to severe economic harm to their states, especially their livestock producers. However, EPA denied these petitions for similar reasons as in 2008.[6]

Another factor that could lead the EPA Administrator to waive certain parts of the RFS is the "blend wall." The vast majority of the RFS is met using ethanol blended into gasoline at the 10% level (E10). However, there is a limit to the amount of ethanol that can be blended into gasoline at this level, and other options for supplying ethanol (e.g., higher-level ethanol blends such as E15 and E85) are currently constrained because of infrastructure and other impediments. In prior years this had not been a problem because the volume of biofuels needed to meet the RFS mandate was below the maximum amount of ethanol that could be supplied as E10; but stakeholders were especially concerned about the possibility of facing this "blend wall" in either 2013 or 2014. In November 2013 EPA proposed using its general authority to lower the RFS mandates in cases of "inadequate supply."[7] EPA proposed an overall mandate of 15.21 billion gallons for 2014, instead of the 18.15 billion gallons of biofuels scheduled for 2014 in EISA, and below the 2013 mandate of 16.55 billion gallons.

Within the overall RFS there are sub-mandates for the use of cellulosic biofuels, biomass-based diesel fuels, and other advanced biofuels. However, questions have been raised over whether there is enough feedstock supply and production capacity to meet some of these carveouts, especially the cellulosic biofuel carveout.[8] EPA waived most of the 2010 cellulosic biofuel carveout—EISA set the mandate at 100 million gallons but EPA only required 6.5 million gallons, more than 90% less than scheduled by EISA. EPA cited a lack of then-current and expected production capacity, driven largely by delays in production plans and a lack of investment in commercial-scale refineries.[9] Similarly, EPA finalized cellulosic biofuel mandates of 6.0 and 10.45 million gallons for 2011 and 2012, respectively—97% and 98% lower than the amounts scheduled in EISA.[10]Because of a lack of commercial cellulosic biofuel production capacity—no cellulosic production was registered in the RFS in 2011 and only about 20,000 gallons were registered in 2012—gasoline and diesel fuel producers challenged the lowered mandates in court. In January 2013, the U.S. Court of Appeals for the D.C. Circuit vacated the 2012 cellulosic mandate and remanded the rule back to EPA.[11] EPA subsequently set the 2012 level at zero and has proposed doing the same for 2011. For 2013, EPA finalized a cellulosic mandate of 6.0 million gallons (instead of 1 billion gallons), and this lowered mandate is also subject to a court challenge.[12] For 2014, EPA has proposed a cellulosic mandate of between 8 and 30 million gallons, instead of 1.75 billion gallons scheduled in the statute.

CURRENT RFS REQUIREMENTS

The RFS requires the use of 16.55 billion gallons of renewable fuel in transportation fuels in 2013—corn ethanol is limited to counting for 13.8 billion gallons of the 2013 mandate. The RFS is scheduled to increase to 36 billion gallons by 2022 with an increasing share coming from "advanced biofuels"—biofuels produced from feedstocks other than corn starch—including cellulosic biofuel and bio-based diesel substitutes. As has been the case in previous years, in 2013 the vast majority of the mandate will be met with U.S. corn ethanol (and a smaller amount of biodiesel, as well as sugarcane ethanol from Brazil). For 2014, the overall mandate is scheduled to increase to 18.15 billion gallons, with 3.75 billion gallons of that coming from advanced biofuels, and 1.75 billion gallons of the advanced biofuel mandate coming from cellulosic biofuels. In November 2013, EPA proposed lowering all three of the mandates, to 15.21 billion gallons, 2.20 billion gallons, and 17 million gallons, respectively.[13]

By 2015 corn ethanol's share of the RFS is effectively capped at 15 billion gallons per year. The EISA amendments to the RFS specifically mandate the use of cellulosic biofuel (16 billion gallons by 2022) and biomass-based diesel fuel (at least 1.0 billion gallons annually by 2012). However, some advanced biofuels, especially cellulosic fuels, have been slow to develop and fuel production lags EISA's mandated schedule.

WAIVER PROVISIONS

As amended by EISA, section 211(o)(7) of the Clean Air Act[14] gives the EPA administrator the authority to waive, in whole or in part, the total volume of renewable fuel mandated by the RFS if, in her determination, there is inadequate domestic supply to meet the mandate, or if "implementation of the requirement would severely harm the economy or environment of a State, a region, or the United States."[15] Further, under certain conditions, the EPA administrator may waive (in whole or in part) the specific carve-outs for cellulosic biofuel and biomass-based diesel fuel.[16]

General Waiver

On the petition of a state or a fuel provider, or at her own discretion,[17] the administrator may waive the overall RFS requirement for a given year. If a waiver is granted, any adjustment applies to total national requirement. Regardless of who initiates the waiver petition, all fuel suppliers' quotas would be reduced by a similar percentage. As the law is written, EPA may not waive the requirement for an individual state or supplier within a state, but must reduce the entire national mandate.

To grant the waiver, the EPA administrator must determine, in consultation with the Secretaries of Agriculture and Energy, that one of two conditions has been met:

- there is inadequate domestic renewable fuel supply; or
- implementation of the requirement would severely harm the economy or environment of a State, a region, or the United States.

However, it is unclear how EPA will interpret these criteria. In its May 1, 2007, final rule for 2007 onward,[18] EPA explicitly stated that it would not establish more specific criteria for the waiver:

> While EPA realizes that the criteria provided by the statute are quite general, the rationales of severe environmental or economic harm or inadequate domestic supply are sufficient for a basic framework upon which a petition can be built and evaluated. Each situation in which a waiver may be requested will be unique, and promulgating a list of more specific criteria in the abstract may be counter-productive.[19]

Within 90 days of receipt of the waiver petition, EPA must act to approve or disapprove the petition, after public notice and opportunity for comment. If EPA does grant a waiver, the waiver expires after one year, but may be extended by the EPA administrator in consultation with the Secretaries of Agriculture and Energy.

Waiver Petitions

In November 2012, EPA denied waiver petitions from the governors of Arkansas, North Carolina, and several other states. Severe droughts across much of the United States in the summer of 2012 which cut into corn yields, along with high worldwide demand for food and feed grains, had pushed December 2012 corn futures prices up from roughly $5.30 per bushel in early

July 2012 to above $8.00 per bushel in mid-August. By mid-September prices had moderated somewhat to the mid-$7.00's.[20] Those high prices raised the feed costs of cattle and other livestock, which may have led to higher consumer meat prices. EPA determined that the RFS requirements alone did not cause severe harm, and a waiver would do little to alleviate any problems in the states. According to EPA,

> it would not be sufficient for EPA to determine that there is severe harm to the economy of a State, region or the United States; rather, EPA must determine that RFS implementation would severely harm the economy. Furthermore, EPA interprets the word "would" as requiring a generally high degree of confidence that implementation of the RFS program would severely harm the economy of a State a region, or the United States. EPA interprets "severely harm" as specifying a high threshold for the nature and degree of harm. Although there are many factors that affect an economy, the RFS waiver provisions call for EPA to evaluate the impact of the RFS mandate itself. EPA does not evaluate the impact of the RFS volume requirements in isolation, but instead evaluates them in the context of all of the relevant circumstances, including in this case the impact of the drought. However the purpose of this analysis is to characterize the impact of the RFS mandate itself, within this context.[21]

Proposed Waiver of 2014 Standards

In November 2013, EPA, under its own authority, proposed lowering the 2014 overall RFS mandate from 18.15 billion gallons to 15.21 billion gallons, and the advanced biofuel mandate from 3.75 billion gallons to 2.20 billion gallons.[22] As noted above, most of the RFS mandate is met through the use of ethanol blended in gasoline at the 10% level (E10), and the volume of gasoline consumed in the United States is basically fixed. Thus, the maximum amount of ethanol that can be supplied in that gasoline is also fixed.[23] Some stakeholders raised concerns that the scheduled volumes for 2014 would exceed the "blend wall" limitations. Others argued that maintaining the levels set in EISA would provide the necessary incentive for the development of infrastructure to provide fuels other than E10.

EPA has proposed using its general authority to lower the RFS mandates in cases of "inadequate supply." In this case, they interpreted "inadequate supply" to include an inability to *deliver* the necessary biofuels to consumers, even if there is sufficient *production* of said biofuels. As noted by EPA,

> [W]e believe that this ambiguous provision is reasonably and best interpreted to encompass the full range of constraints that could result in

an inadequate supply of renewable fuel to the ultimate consumers, including fuel infrastructure and other constraints. This would include, for instance, factors affecting the ability to produce or import qualifying renewable fuels as well as factors affecting the ability to distribute, blend, dispense, and consume those renewable fuels.

The waiver provision at CAA 211(o)(7)(A)(ii) is ambiguous in several respects. First, it does not specify what the general term "supply" refers to. The common understanding of this term is an amount of a resource or product that is available for use by the person or place at issue. [footnote omitted] Hence the evaluation of the supply of renewable fuel, a product, is best understood in terms of the person or place using the product. In the RFS program, various parties interact across several industries to drive the ultimate use of renewable fuel by consumers of transportation fuel. For example, supplying renewable fuel to obligated parties and terminal blenders is one part of this process, while supplying renewable fuel to the ultimate consumer as part of transportation fuel is a different and later aspect of this process. This is clearly the case with respect to the renewable fuels ethanol and biodiesel, which are typically supplied to the obligated parties and terminals as a neat fuel, but in almost all cases are supplied to the consumer as a blend with conventional fuel (ethanol and gasoline or biodiesel and diesel). The waiver provision does not specify what product is at issue (for example, neat renewable fuel or blended renewable fuel with transportation fuel) or the person or place at issue (for example, obligated party or ultimate consumer), in determining whether there is an "inadequate domestic supply."[24]

Gasoline and diesel fuel suppliers and other stakeholders have generally been supportive of EPA's proposal,[25] while many biofuel producers believe that EPA is overstepping its authority.[26]

Cellulosic Biofuel Waiver[27]

As part of the RFS, EISA established a specific mandate for the use of cellulosic biofuels— ethanol or other fuels produced from woody or fibrous materials such as grasses, trees, etc. The cellulosic carveout was to start in 2010 at 100 million gallons, and increase to 16.0 billion gallons by 2022. Current cellulosic biofuel production is limited, with few commercial-scale plants in operation. Because of uncertainties over production capacity and cellulosic biofuel supply, in its February 2010 rulemaking, EPA reduced the cellulosic mandate from 100 million gallons to 6.5 million gallons for 2010.[28] EPA set cellulosic mandates of 6.0 million gallons, 10.45 million gallons, and

6.0 million gallons for 2011, 2012, and 2013 respectively, well below the scheduled amounts of 250 million gallons, 500 million gallons, and 1.0 billion gallons for those years. EPA has proposed a 2014 standard of 17 million gallons, significantly lower than the scheduled level of 1.75 billion gallons.

If the EPA administrator determines that the projected production volume of cellulosic biofuel for a given year is less than the mandated amount, she may reduce the carve-out. If she is going to do so, the administrator must reduce the required amount by November 30 of the preceding calendar year. If the administrator does reduce the mandated amount of cellulosic biofuel, she may also reduce the required amount of advanced biofuel as well as the total volume required for that year under the RFS by an equal or lesser amount, but she is *not required* to do so. For 2010 through 2013, EPA retained the overall RFS and advanced biofuel levels mandated in EISA.[29] For 2014, EPA has determined that there would be insufficient advanced biofuel supply to make up the shortfall in cellulosic supply, and has proposed lowering both the advanced biofuel and overall RFS.

Unlike the general waiver, only the EPA administrator may initiate a decision on a cellulosic biofuel waiver (as opposed to, e.g., a state). Because of a lack of commercial cellulosic biofuel production capacity—no cellulosic production was registered in the RFS in 2011 and only about 20,000 gallons were registered in 2012—gasoline and diesel fuel producers challenged the lowered mandates in court. In January 2013, the U.S. Court of Appeals for the D.C. Circuit vacated the 2012 cellulosic mandate and remanded the rule back to EPA.[30] EPA subsequently set the 2012 level at zero and has proposed doing the same for 2011.

Biomass-Based Diesel Waiver

Similar to the cellulosic biofuel carve-out, EISA also established a specific mandate for the use of biomass-based diesel (BBD) fuel. Currently, most of this fuel is "biodiesel"—a diesel fuel substitute produced from soybean oil and other vegetable oils through a process called "transesterification"—but other fuels, some of which are termed "renewable diesel," would also qualify. The BBD carve-out started in 2009 at 0.5 billion gallons and increases to a minimum of 1.0 billion gallons by 2012. Because the EPA's rule was not finalized until 2010, the Agency established a combined biomass diesel mandate for 2009 and 2010 of 1.15 billion gallons.[31] For 2012, the BBD carveout is set at 1.0 billion gallons. Because there is domestic capacity to produce BBD beyond the 1.0 billion gallon floor, EPA finalized a 2013 BBD mandate of 1.28 billion gallons,[32] and has proposed the same level for 2014 and 2015.[33]

If the EPA administrator (in consultation with the Secretaries of Energy and Agriculture) determines that there are significant market circumstances (including feedstock disruptions) "that would make the price of BBD fuel increase significantly," the administrator may reduce the amount mandated for up to 60 days.[34] The administrator may extend the waiver for no more than an additional 60 days. To date EPA has not exercised this authority. It should be noted that tax credits that have supported the development of biomass-based diesel fuels expired at the end of 2013.[35] Without the credits, it is unclear how much BBD will be produced in 2014. It is possible that the expiration of the credit could constrain BBD supply in 2014 or later years. The tax credits have expired and been extended (sometimes retroactively) several times.

Reductions in the RFS

Starting in 2016, if the administrator waives a significant share of the above requirements, she must reduce the required volumes in all subsequent years. Specifically, she must reduce the applicable amounts in future years if she waives any of the above requirements by

- 20% or more for two consecutive years; or
- 50% or more in a single year.

For example, if the administrator reduced the overall RFS requirement by 6.0 billion gallons in both 2017 and 2018, then she would be required to reduce the total RFS requirement by 6.0 billion gallons in 2019 and beyond. The one exception is that these reductions in the RFS would not apply to the requirements before calendar year 2016.

Effects of a Waiver

Questions have been raised over how a waiver approval would affect food and fuel markets. As these markets are extremely complex, there is no simple answer. The effects of a waiver would likely depend on many factors:

- the degree to which the RFS requirements are relaxed under the waiver;
- the duration of the waiver;
- the scope of the waiver (cellulosic biofuel, biomass-based diesel, or the entire program);
- whether the waiver is extended; and
- prevailing supply and prices for oil, gasoline, biofuels, and feedstock commodities.

In the specific case of the 2010 waiver for cellulosic biofuel, a key question was whether this waiver would undermine the credibility of the mandates in future years and undercut investment. As EPA noted in the final rule establishing the expanded RFS, "In the proposal, we did a preliminary assessment of the cellulosic biofuel industry to arrive at the conclusion that it was possible to uphold the 100 million gallon standard in 2010 based on anticipated production."[36]

In the final rule on the RFS expansion, EPA did not state whether or not the agency believed there would be sufficient capacity to meet the cellulosic mandates in coming years (250 million gallons in 2011, 500 million gallons in 2012, and 1 billion gallons in 2013), although EPA stated that "it is remarkable how much progress the industry has made in such a short time, and there is a tremendous growth opportunity for cellulosic biofuels over the next several years."[37] Ultimately, for 2010 thorough 2013, EPA found that there was insufficient production capacity to meet the scheduled levels of cellulosic biofuel, a situation likely to be continued in 2014. If EPA continues to find that mandates in later years likewise are unachievable, and if investors assume that future waivers are unavoidable, further investment in cellulosic biofuel refineries may be limited. Similarly, EPA's proposal to partially waive the overall RFS requirements for 2014 may hamper investment in biofuels production capacity. On the other hand, requiring smaller amounts of biofuels in 2014 may put downward pressure on corn demand and corn prices, which could affect many other markets.

End Notes

[1] For more information on the RFS, see CRS Report R40155, Renewable Fuel Standard (RFS): Overview and Issues, by Randy Schnepf and Brent D. Yacobucci, and CRS Report R43325, The Renewable Fuel Standard: In Brief, by Kelsi Bracmort.

[2] Rick Perry, governor of Texas, Letter to The Honorable Stephen L. Johnson, Administrator, U.S. Environmental Protection Agency, April 25, 2008.

[3] U.S. Environmental Protection Agency, EPA Decision on Texas Request for Waiver of Portion of Renewable Fuel Standard (RFS), EPA420-F-08-029, Washington, DC, August, 2008, http://www.epa.gov/otaq/renewablefuels/ 420f08029.htm.

[4] Letter from Mike Beebe, Governor of Arkansas, to Lisa P. Jackson, EPA Administrator, August 13, 2012, http://www.epa.gov/otaq/fuels/renewablefuels/documents/arkansas-rfs-waiver-request.pdf. Letter from Beverly Eaves Perdue, Governor of North Carolina, to Lisa P. Jackson, EPA Administrator, August 14, 2012, http://www.epa.gov/ otaq/fuels/ renewablefuels/documents/north-carolina-rfs-waiver-request.pdf.

[5] New Mexico, Georgia, Texas, Virginia, Maryland, Delaware, Utah, and Wyoming. The governor of Florida wrote EPA in support of the waiver requests. 77 Federal Register 70754.

[6] U.S. Environmental Protection Agency, "Notice of a Decision Regarding Requests for a Waiver of the Renewable Fuel Standard," 77 Federal Register 70752-70776, November 27, 2012.

[7] U.S. Environmental Protection Agency, "2014 Standards for the Renewable Fuel Standard Program; Proposed Rule," 78 Federal Register 71732-71784, November 29, 2013.

[8] For more information on cellulosic biofuels, see CRS Report R41106, Meeting the Renewable Fuel Standard (RFS) Mandate for Cellulosic Biofuels: Questions and Answers, by Kelsi Bracmort.

[9] U.S. Environmental Protection Agency, Regulation of Fuels and Fuel Additives: Changes to Renewable Fuel Standard Program; Final Rule, EPA-HQ-OAR-2005-0161, Washington, DC, February 3, 2010, pp. 173-174, http://www.epa.gov/otaq/renewablefuels/rfs2-preamble.pdf.

[10] Environmental Protection Agency, "Regulation of Fuel and Fuel Additives: 2011 Renewable Fuel Standards; Final Rule," 75 Federal Register 76790, December 9, 2010; and Environmental Protection Agency, "Regulation of Fuels and Fuel Additives: 2012 Renewable Fuel Standards; Final Rule," 77 Federal Register 1320, January 9, 2012.

[11] API v. EPA, 706 F.3d 474 (D.C. Cir. 2013).

[12] Erin Voegele, "API Challenges 2013 RFS Cellulsoic Volume Requirement," Biomass Magazine, October 8, 2013.

[13] U.S. Environmental Protection Agency, "2014 Standards for the Renewable Fuel Standard Program; Proposed Rule," 78 Federal Register 71732-71784, November 29, 2013.

[14] 42 U.S.C. 7545(o)(7).

[15] 42 U.S.C. 7545(o)(7)(A)(i).

[16] 42 U.S.C. 7545(o)(7)(D)-(E).

[17] Under EPAct, only states could petition EPA for the waiver. EISA amended the RFS to allow fuel providers to file a petition, and to give the EPA administrator authority to initiate the process on her own motion.

[18] EPA, Regulation of Fuels and Fuel Additives: Renewable Fuel Standard Program; Final Rule, May 1, 2007. 72 Federal Register 23899-24014.

[19] 72 Federal Register 23928.

[20] CME Group, Corn Futures, September 24, 2012, http://www.cmegroup.com/trading/ agricultural/grain-and-oilseed/ corn.html.

[21] U.S. Environmental Protection Agency, "Notice of a Decision Regarding Requests for a Waiver of the Renewable Fuel Standard," 77 Federal Register 70756, November 27, 2012.

[22] U.S. Environmental Protection Agency, "2014 Standards for the Renewable Fuel Standard Program; Proposed Rule," 78 Federal Register 71732-71784, November 29, 2013.

[23] For more information, see CRS Report R40445, Intermediate-Level Blends of Ethanol in Gasoline, and the Ethanol "Blend Wall", by Brent D. Yacobucci.

[24] U.S. Environmental Protection Agency, "2014 Standards for the Renewable Fuel Standard Program; Proposed Rule," 78 Federal Register 71755-71756, November 29, 2013.

[25] American Petroleum Institute, Industry, Consumer Groups Say EPA Took Steps to Protect Consumers from Blend Wall, but More Needs to Be Done, Washington, DC, November 2013, http://www.api.org/news-and-media/news/ newsitems/2013/nov-2013/epa-took-steps-to-protect-consumers-from-blend-wall-but-more-needs-to-be-done.

[26] Holly Jessen, "Examining EPA's Authority to Reduce RFS Volume Requirements," Ethanol Producer Magazine, November 22, 2013, http://www.ethanolproducer.com/articles/ 10489/ examining-epas-authority-to-reduce-rfs-volumerequirements.

[27] For more information on cellulosic fuels, see CRS Report R41106, Meeting the Renewable Fuel Standard (RFS) Mandate for Cellulosic Biofuels: Questions and Answers, by Kelsi Bracmort.

[28] For example, EPA cited projections from the Energy Information Administration (EIA) that roughly 5 million gallons of cellulosic fuels could be produced in 2010, although some of that fuel would be cellulosic diesel fuel, which generates more credits per gallon due to its higher energy content. As finalized in the rule, the RFS requirements are based on ethanol-equivalent gallons, and the 5 million gallon number from EIA translates to roughly 6.5 million ethanol-equivalent gallons. U.S. Environmental Protection Agency, Regulation of Fuels and Fuel Additives: Changes to Renewable Fuel Standard Program; Final Rule, EPA-HQ-OAR-2005-0161, Washington, DC, February 3, 2010, p. 174, http://www.epa.gov/ otaq/renewablefuels/rfs2-preamble.pdf.

[29] In general, the excess amount of advanced biofuel required because of the cellulosic waviers has been met using biomass-based diesel in excess of the biomass-based diesel carveout, as well as other fuels defined as advanced biofuels under EISA, including ethanol produced from Brazilian sugarcane.

[30] API v. EPA, 706 F.3d 474 (D.C. Cir. 2013).

[31] In this way, biomass-based diesel credits generated in 2009 can be used for compliance in 2009-2010. If the rule applied only to 2010, those credits generated by fuel blenders in 2009 would have been useless.

[32] Environmental Protection Agency, Regulation of Fuels and Fuel Additives: 2013 Biomass-Based Diesel Renewable Fuel Volume, Prepublication Version, Washington, DC, September 14, 2012, http://www.epa.gov/otaq/fuels/ renewablefuels/documents/rfs-biomass-diesel-std-fr.pdf.

[33] U.S. Environmental Protection Agency, "2014 Standards for the Renewable Fuel Standard Program; Proposed Rule," 78 Federal Register 71737, November 29, 2013.

[34] However, the amount may not represent more than 15% of the total required amount for that year.

[35] For more information on BBD tax credits and other biofuels incentives, see CRS Report R42566, Alternative Fuel and Advanced Vehicle Technology Incentives: A Summary of Federal Programs, by Lynn J. Cunningham et al.

[36] U.S. Environmental Protection Agency, Regulation of Fuels and Fuel Additives: Changes to Renewable Fuel Standard Program; Final Rule, EPA-HQ-OAR-2005-0161, Washington, DC, February 3, 2010, p. 173, http://www.epa.gov/otaq/renewablefuels/rfs2-preamble.pdf.

[37] Ibid., p. 178.

In: The Renewable Fuel Standard
Editor: Lydia Rider
ISBN: 978-1-63321-829-1

Chapter 3

ANALYSIS OF RENEWABLE IDENTIFICATION NUMBERS (RINS) IN THE RENEWABLE FUEL STANDARD (RFS)*

Brent D. Yacobucci

SUMMARY

The federal Renewable Fuel Standard (RFS) was established in the Energy Policy Act of 2005 (EPAct) and significantly expanded in the Energy Independence and Security Act of 2007 (EISA). The RFS requires the use of renewable biofuels in transportation fuel. For 2013, the RFS requires the use of 16.55 billion gallons of renewable fuel. Within the larger mandate, there are sub-mandates ("carve-outs") for advanced biofuels (e.g., biomass-based diesel and cellulosic fuels). By 2022, the RFS requires the use of 36 billion gallons of renewable fuels, including 21 billion gallons of advanced biofuels.

The RFS is a market-based compliance system in which obligated parties (generally refiners and/or terminal operators) must submit credits to cover their obligations. These credits— Renewable Identification Numbers, or RINs—are effectively commodities that can be bought or sold like other commodities. For each gallon of renewable fuel in the RFS program, one RIN is generated. Each RIN is a 38-digit number, with

* This is an edited, reformatted and augmented version of a Congressional Research Service publication R42824, prepared for Members and Committees of Congress, dated July 22, 2013.

blocks of digits corresponding to various data, including the year the RIN was generated, the producer of the fuel, and the type of fuel. RINs are valid for use in the year they are generated and the following year.

From the beginning of the RFS program, there have been concerns with RIN generation and the RIN market. As the RINs are essentially numbers in a computerized account, there have been errors and opportunities for fraud. Because of concerns over transposed digits, invalid characters, allegations of double-counting (intentional or unintentional) and other errors and inaccuracies, when EPA finalized rules for the RFS as expanded by EISA (the "RFS2"), EPA also established a new in-house trading system in an effort to address these concerns. All RIN transactions must be cleared through this in-house system, called the EPA Moderated Transaction System (EMTS). From the beginning of the RFS2 EPA has maintained that all due diligence remains the duty of obligated parties. Under this "buyer beware" system those purchasing or receiving RINs must certify their validity on their own, and they are responsible for any fraudulent RINs they pass on to other buyers or submit to EPA for compliance.

In late 2011 and early 2012, EPA issued Notices of Violations (NOVs) to three companies that the agency alleges fraudulently generated a combined 140 million biodiesel RINs in 2010 and 2011. Because of these RIN fraud cases, EPA is looking at establishing a system whereby RINs can be certified by third parties registered with EPA. EPA is considering whether such certification would provide obligated parties with an "affirmative defense" if RINs are later found to fraudulent—that is, obligated parties would not be liable for penalties under the Clean Air Act for the use of such RINs. Key questions include whether such an affirmative defense would also eliminate the requirement to purchase make-up RINs. EPA proposed a plan in January 2013, and projects that the rule will be finalized in October 2013.

Most RINs are bought and sold through private contracts registered with the EMTS. However, there are also spot markets for RINs, and since the beginning of 2013, conventional ethanol RIN prices have risen dramatically. Prices rose from roughly $0.07 per gallon in early January to over $1.00 per gallon in early July. Various factors have been identified by stakeholders as potentially causing the price increase, including whether sufficient amounts of ethanol can be blended into gasoline to meet the 2013 RFS mandates and the extent to which non-obligated parties are speculating in RIN markets.

In the 113th Congress several congressional hearings have been held and various bills have been proposed to address both RIN issues and the overall RFS.

INTRODUCTION

The Energy Policy Act of 2005 (EPAct, P.L. 109-58) established a renewable fuel standard (RFS), requiring the use of biofuels (such as ethanol) in the nation's fuel supply. The Energy Independence and Security Act of 2007 (EISA, P.L. 110-140) significantly expanded this mandate.[1] The RFS mandate has been a major impetus to the development of U.S. biofuels industries, especially the ethanol and biodiesel industries. In 2005, the United States produced 3.9 billion gallons of ethanol and 0.1 billion gallons of biodiesel. In 2011, production had increased to roughly 14 billion gallons of ethanol and 1 billion gallons of biodiesel.

Covered parties meet their obligations under the RFS by surrendering renewable fuel credits to EPA equal to the number of gallons in their annual obligation. These credits, known as Renewable Identification Numbers (RINs), are generated when a batch of biofuel is produced, and separated from the fuel by obligated parties (generally gasoline and diesel fuel refiners or blenders). Once separated, these RINs may be traded like other commodities. Recent civil and criminal action against parties accused of registering and selling fraudulent RINs has raised questions about the integrity of the RIN market and EPA's oversight of the market.

This report outlines the RFS and the current RIN system, discusses the current market for various RINs, and outlines policy considerations to address RIN fraud going forward.

CURRENT RFS REQUIREMENTS

For 2013, the RFS requires the blending of 16.55 billion gallons of renewable fuel in transportation fuels, including at least 1.28 billion gallons of biomass-based diesel substitutes (BBD). The RFS increases to 36 billion gallons by 2022 with an increasing share coming from "advanced biofuels"—biofuels produced from feedstocks other than corn starch—including cellulosic biofuel and BBD fuels. As has been the case in previous years, in 2013 the vast majority of the mandate is expected to be met with U.S. corn ethanol (and a smaller amount of biodiesel, as well as sugarcane ethanol from Brazil).

By 2015 corn ethanol's share of the RFS is effectively capped at 15 billion gallons per year. The EISA amendments to the RFS specifically mandate the use of cellulosic biofuel (16 billion gallons by 2022) and biomass-based diesel fuel (at least 1.0 billion gallons annually by 2012). However, advanced biofuels, especially cellulosic fuels, have been slow to develop and fuel production lags the EISA's mandate schedule.[2]

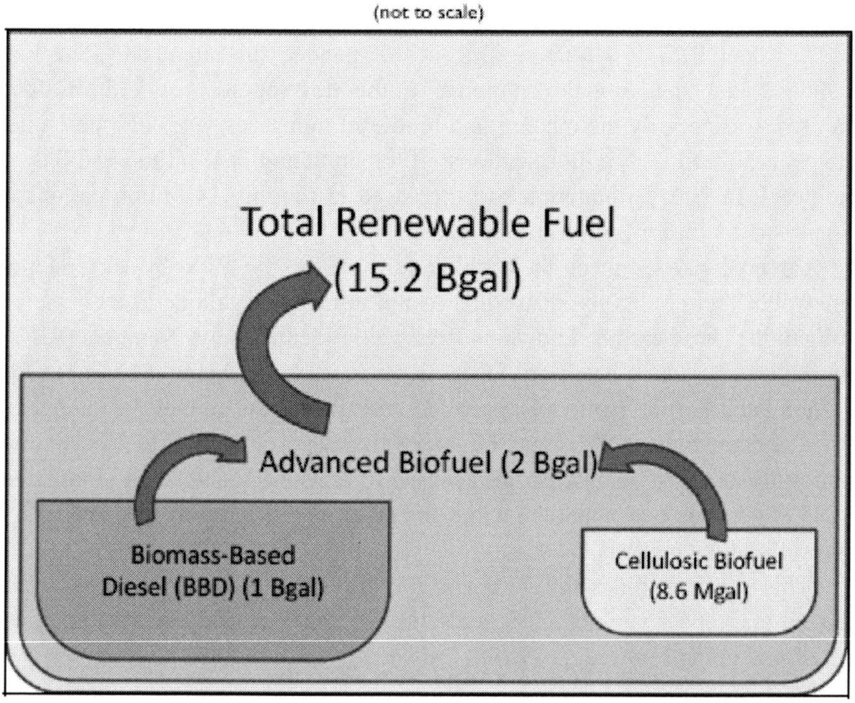

Source: CRS.

Notes: As noted by the arrows, fuel qualifying as one type of biofuel in the RFS qualifies for all levels above it. For example, cellulosic biofuel may also be used to meet the advanced biofuel mandate and the overall RFS mandate. However, non-cellulosic advanced biofuel (e.g., sugarcane ethanol) may not be used to meet the cellulosic or BBD mandates. Likewise, corn starch ethanol may only be used to meet the total RFS mandate (and not the advanced, cellulosic, or BBD mandates). For illustrative purposes only, an 8.6 million gallons value is used for cellulosic biofuels. The 8.6 million physical gallons (10.45 million gallons ethanol-equivalent) was the required amount set by EPA before a federal court vacated the 2012 cellulosic biofuel standard. Following the court decision, EPA dropped the 2012 RFS for cellulosic biofuels to zero.

Figure 1. Nested RFS Mandates for 2012.

Within the overall RFS mandate, there are sub-mandates for specific types of fuel. For example, for 2012 EISA required the use of 15.2 billion gallons of biofuels, of which 2.0 billion must be "advanced biofuels." Within the advanced biofuel carve-out for 2012, at least 1.0 billion gallons were to be biomass-based diesel (BBD) fuels and zero gallons were required for cellulosic biofuels.[3] In the early years of the program, the lion's share of the mandate is unspecified, and the vast majority of this unspecified portion has been—and is expected to be—supplied by corn-based ethanol largely produced in the Midwest. At the beginning of each year, EPA determines a percentage standard that all suppliers must meet, which is based on expected total U.S. gasoline and diesel demand for the prior year. For example, for 2012, the overall biofuel standard was 9.23%, the advanced biofuel standard was 1.21%, the BBD standard was 0.91%, and the cellulosic biofuel standard was 0.006%.[4]

The sub-mandates for advanced biofuels are nested together (**Figure 1**). As noted by the arrows in the figure, fuel qualifying as one type of biofuel in the RFS qualifies for all levels above it. For example, a gallon of cellulosic biofuel may be used to meet the cellulosic mandate, the advanced biofuel mandate, and the overall RFS. A gallon of other advanced biofuel (e.g., sugarcane ethanol) may be used to meet the advanced biofuel mandate and the overall mandate, but may not be used to meet the cellulosic or BBD mandates. Corn starch ethanol—the most widely used biofuel in the United States—may only be used to meet the overall RFS.[5]

THE ROLE OF RINS

Compliance with the RFS is measured using RINs. When qualifying biofuels are produced, each gallon is assigned a RIN. Until the biofuels are sold as fuel or blended into conventional fuels, the RINs are "attached" to the fuel. Once the biofuel has been blended or sold, the RINs are detached, and can then be bought and sold like other commodities. At the end of each year, fuel suppliers must multiply the above percentage standards by their total gasoline and diesel sales to calculate their renewable volume obligations (RVO), which indicate the total number of each type of RIN that the suppliers must submit to EPA. To the extent that a supplier has excess RINs, that supplier may sell them to others who may be short, or save them for use in the following year.

RINs

A RIN is a unique 38-character number that is issued (in accordance with EPA guidelines) by the biofuel producer or importer at the point of biofuel production or the port of importation.[6] Each qualifying gallon of renewable fuel has its own unique RIN. RINs are generally assigned by batches of renewable fuel production. (See box at right.)

Under the RFS2 RIN formulation, Code D identifies which of the four RFS categories—total, advanced, cellulosic, or biodiesel—the biofuel satisfies. Together, SSSSSSSS and EEEEEEEE identify the RIN block which demarcates the number of gallons of renewable fuel that the batch represents in the context of compliance with the RFS—that is, RIN gallons. The RIN-gallon total equals the product of the liquid volume of renewable fuel times its energy equivalence value (relative to a gallon of ethanol). For example, because biodiesel has an equivalence value (EV) of 1.5 when being used as an advanced biofuel, 1,000 gallons of biodiesel would equal 1,500 RIN gallons of advanced biofuels.[7] If the RIN block start for that batch was 1 (i.e., SSSSSSSS = 00000001), then the end value (EEEEEEEE) would be 00001000.

RIN Codes

RIN=KYYYYCCCCFFFFFBBBBBRRDSSSSSSSSEEEEEEEE

Where

K = code distinguishing RINs still assigned to a gallon from RINs already separated

YYYY = the calendar year of production or import

CCCC = the company ID

FFFFF = the company plant or facility ID

BBBBB = the batch number

RR = the biofuel energy equivalence value

D = the renewable fuel category

SSSSSSSS = the start number for this batch of biofuel

EEEEEEEE= the end number for this batch of biofuel

Any party that owns RINs at any point during the year (including domestic and foreign producers; refiners and blenders; exporters and importers of renewable fuels; and RIN traders) must register with the EPA and follow RIN record-keeping and reporting guidelines. RINs can only be generated if it can be established that the feedstock from which the fuel was made meets EISA's definitions of renewable biomass (including land-use restrictions), and if the fuel meets EISA's lifecycle greenhouse gas emission limits. The feedstock affirmation and record-keeping requirements apply to RINs generated by both domestic renewable fuel producers and RINgenerating foreign renewable fuel producers or importers.

EPA Moderated Transaction System (EMTS)

All RIN transactions, including generation, trade/sale/transfer, separation, and retirement, must be cleared through the EMTS. When biofuels change ownership (e.g., are sold by a producer to a blender), any attached RINs are also transferred.[8] The Code K status of the RIN is changed at separation (generally when the fuel is sold from a biofuel producer to an obligated party). (See *Figure 2*.)

As noted by EPA in the rule establishing the RFS2 and the EMTS, EPA views the EMTS solely as a "screening" system, and all due diligence remains the duty of obligated parties.[9] Under this "buyer beware"[10] system those purchasing or receiving RINs must certify their validity on their own, and they are responsible for any fraudulent RINs they pass on to other buyers or submit to EPA for compliance.

THE MARKET FOR RINS

RIN Prices

Because RINs may be bought and sold as commodities, there are RIN spot markets. However, these spot markets may only provide some insight into the actual value of the total pool of RINs in a given year, as RINs may or may not be traded after they are detached by fuel suppliers. Because RINs are not completely fungible, their values may or may not be affected by the markets for other RINs. For example, RINs for conventional ethanol may only be used for the overall (unspecified) renewable fuel mandate. However, biodiesel RINs may be used to meet the BBD, advanced biofuel, and overall RVOs.

It should also be noted that unlike other commodities, RINs generally may only be used in the year they are generated or for one additional year, although suppliers may only meet up to 20% of their current-year obligation with the previous year's RINs. Thus, RIN values diminish over time and ultimately have no value in the second year after they are generated.

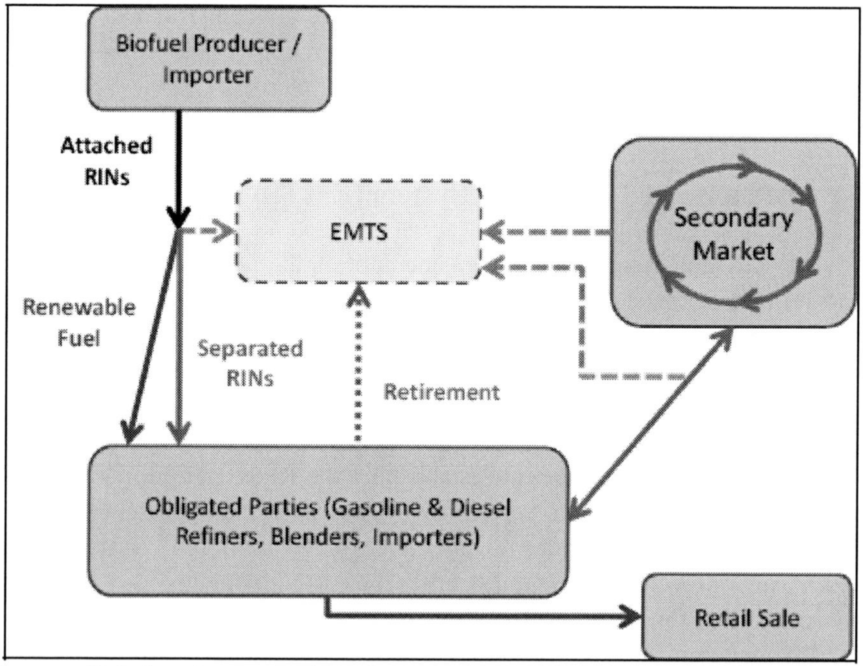

Source: CRS, based on Rakesh Radhakrishnan, Market Considerations—RECs and RINs Overlap, Thompson Reuters, September 25, 2012, p. 8, http://www. renewableenergymarkets.com/docs/ presentations/2012/Radhakrishnan.pdf.

Notes: Black lines indicate RINs attached to actual biofuel gallons. Solid blue lines indicate separated RINs that may be traded among all market participants. Dashed blue line indicates end-of-year submission of RINs by obligated parties to EPA to meet RFS mandates. Green lines indicate actual biofuel gallons separated from RINs. Orange lines indicate that all RIN transactions must be cleared through EMTS.

Figure 2. Simplified Schematic of RIN Trading System.

Prices Through 2012

Through 2012 there was generally an excess of corn ethanol (beyond what was allowed for meeting the unspecified portion of the RFS) in the U.S. market, and ethanol RINs generally traded at much lower prices than other RINs—generally between one and four cents per gallon, as opposed to a dollar per gallon or more for other fuels. (See *Figure 3*, *Figure 4*, and *Figure 5*.) Because much of the advanced biofuel mandate is met using BBD RINs, the advanced biofuel RIN price follows the BBD RIN price closely when adjusted for energy content. For example, when BBD RIN prices spiked in September 2011 (*Figure 4*), advanced biofuel RINs showed a similar spike (*Figure 5*); there was no spike at that time in corn ethanol RINs (*Figure 3*). Similarly, BBD and advanced biofuel RIN prices dropped in the second half of 2012 (apparently driven by concerns over the validity of biodiesel RINs), while ethanol RIN prices increased (but still remained relatively low) as the 2012 drought raised concerns over U.S. corn production and its effects on ethanol production.

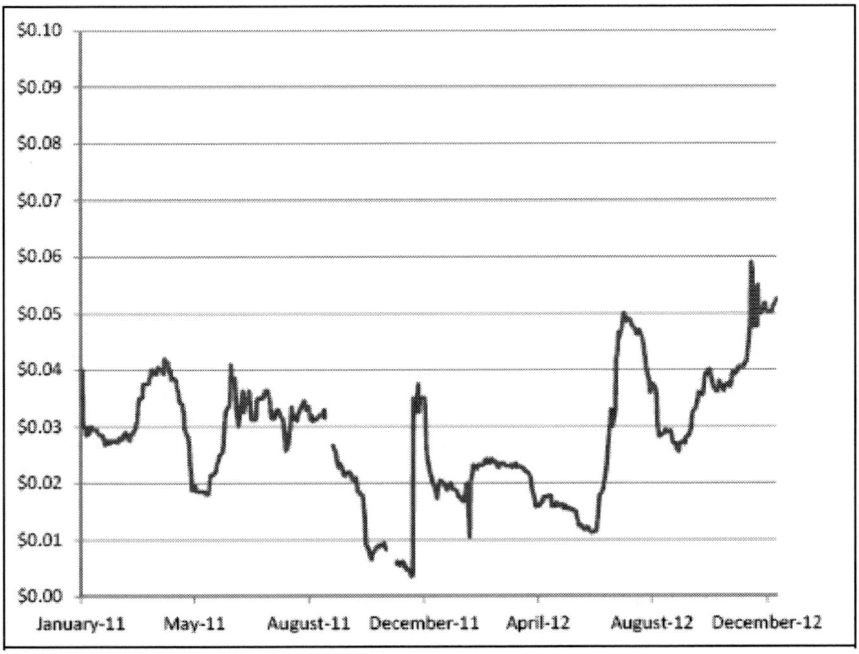

Source: "Ethanol and Gasoline Component Spot Market Prices," OPIS Ethanol & Biodiesel Information Service, various editions (January 10, 2011-January 7, 2013).

Notes: Average of daily high and low prices reported by OPIS. Most biofuels are sold under contract, and thus spot prices may not reflect the value of all RINs traded at any given time.

Figure 3. Spot Renewable Fuel (Corn Ethanol) RIN Prices, 2011-2012.

Source: "Ethanol and Gasoline Component Spot Market Prices," OPIS Ethanol & Biodiesel Information Service, various editions (January 10, 2011-January 7, 2013).

Notes: Average of daily high and low prices reported by OPIS. Most biofuels are sold under contract, and thus spot prices may not reflect the value of all RINs traded at any given time.

Figure 4. Spot BBD RIN Prices, 2011-2012.

Prices Since January 2013

Spot prices for conventional (corn-based) ethanol RINs have risen dramatically since the start of 2013. On January 1, ethanol RINs were trading at roughly $0.07 per gallon, but they spiked to over $1.00 in mid-March, before receding and then rising to around $1.16 by July 11 (*Figure 6*). Several factors have been identified by stakeholders as potential causes for the run-up in prices. For example, many stakeholders are concerned that the fuel supply is rapidly approaching a "blend wall."[11] There are various factors that limit ethanol content in gasoline to a maximum of 10%. With a limited number of gasoline gallons sold each year, but an increasing RFS mandate, the amount of ethanol needed to meet the mandate may exceed the potential for gasoline blending in the near future.

Source: "Ethanol and Gasoline Component Spot Market Prices," OPIS Ethanol & Biodiesel Information Service, various editions (January 10, 2011-January 7, 2013).

Notes: Average of daily high and low prices reported by OPIS. Most biofuels are sold under contract, and thus spot prices may not reflect the value of all RINs traded at any given time.

Figure 5. Spot Advanced Biofuel RIN Prices, 2011-2012.

To address the blend wall and still meet the RFS mandates, obligated parties will need to do one of the following: (1) blend ethanol in gasoline at higher concentrations (e.g., 15%); (2) sell ethanol as an alternative fuel for flexible fuel vehicles (FFVs); or (3) supply non-ethanol biofuels. If none of these options are undertaken in sufficient quantity, there may be insufficient RINs for obligated parties to meet their RVOs. The rise in spot prices for corn ethanol RINs in 2013 may reflect a scarcity in RINs not seen in previous years.

A second concern some stakeholders have raised is that non-fuel traders may be speculating in RIN markets and pushing up prices.[12]

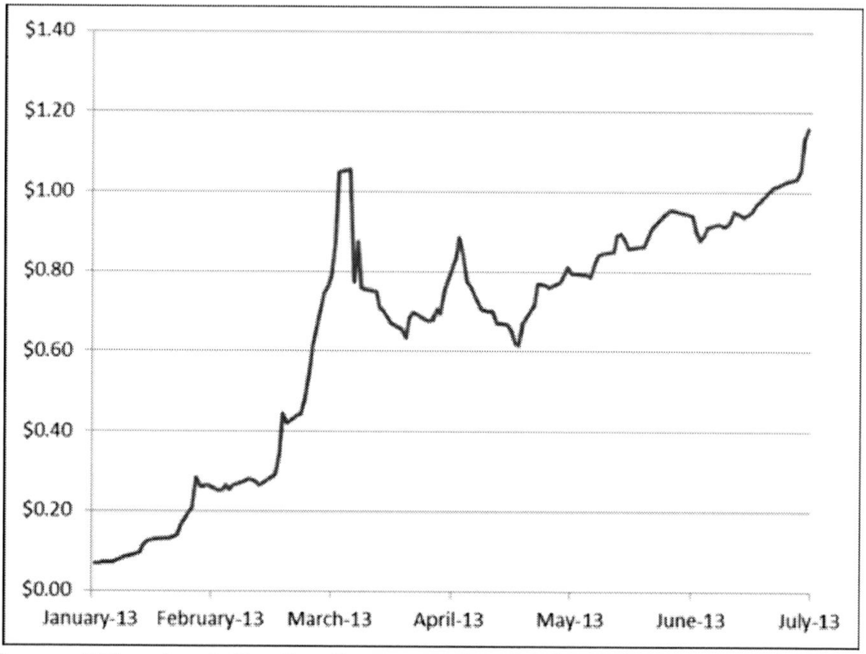

Source: "Ethanol and Gasoline Component Spot Market Prices," OPIS Ethanol & Biodiesel Information Service, various editions (January 7, 2013-July 15, 2013).

Notes: Average of daily high and low prices reported by OPIS. Most biofuels are sold under contract, and thus spot prices may not reflect the value of all RINs traded at any given time.

Figure 6. Spot Renewable Fuel (Corn Ethanol) RIN Prices. (January to July 11, 2013)

RIN Volumes

The market for RINs is potentially very large, although the amount of RIN trading that occurs is unclear. Although EPA reports total RINs registered by month, and the EMTS tracks trades and RIN prices, EPA does not report these data. Likewise, publicly available data from other sources are similarly limited.[13] *Figure 7* and *Figure 8* show total RINs registered for 2011 and 2012. As noted above, by volume the RFS is dominated by ethanol produced from corn starch.

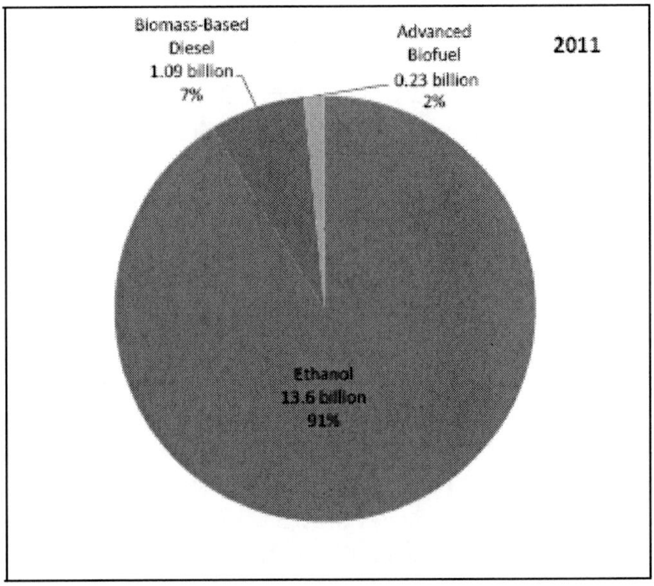

Source: EPA, EPA Moderated Transaction System (EMTS).

Figure 7. Total RINs Registered 2011.

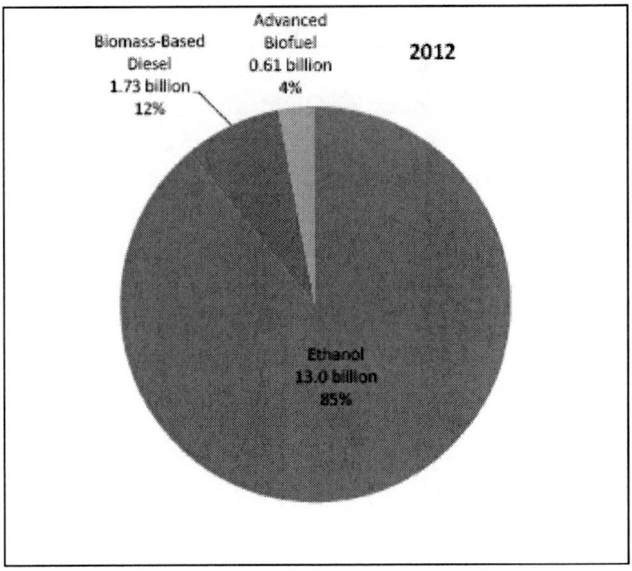

Source: EPA, EPA Moderated Transaction System (EMTS).

Figure 8. Total RINs Registered 2012.

FRAUDULENT RINS

As noted above, in late 2011 and early 2012, EPA issued Notices of Violations (NOVs) to three companies (Clean Green Fuels, LLC, Absolute Fuels, LLC, and Green Diesel, LLC) that the agency alleges fraudulently generated a combined 140 million biodiesel RINs in 2010 and 2011.[14] Subsequently, individuals representing two of these companies have also faced criminal prosecution.[15] Because these investigations involve potentially criminal actions, EPA has limited the amount of information available to traders and obligated parties who may have purchased fraudulent RINs. Thus, it is unclear whether any other NOVs will be issued in the future.

The 140 million fraudulent RINs from the three NOVs represent roughly 11% of the biodiesel RINs generated between mid-2010 and the end of 2011, but less than 1% of the total RINs generated—as noted above, ethanol produced from corn starch currently dominates the RFS. However, biodiesel RINs traded at considerably higher prices than ethanol RINs at that time.

Effects on Obligated Parties

In the regulations establishing the RFS2 and the EMTS, EPA specifically stated that "invalid RINs cannot be used to achieve compliance with the Renewable Volume Obligations (RVO) of an obligated party or exporter, regardless of the party's good faith belief that the RINs were valid at the time they were acquired."[16] Because of the "buyer beware" nature of the system, obligated parties who purchased the fraudulent RINs must pay fines for each RIN submitted (EPA and the companies have generally settled at about $0.10 per RIN), and must submit valid RINs to offset the fraudulent RINs. Thus, the combined economic costs to the obligated parties may include:

1. the original cost of the fraudulent RINs (spot prices ranged between $0.70 and $2.00 per RIN over that time);
2. penalties to EPA for Clean Air Act violations ($0.10 per RIN, capped at $350,000 per party);
3. the cost of all make-up RINs (trading at the time of settlement at roughly $0.50 per gallon); and
4. any legal costs in pursuing restitution from fraudulent actors.

Quality Assurance Program

Because of these RIN fraud cases, EPA is looking at establishing a quality assurance program whereby RINs can be certified by third parties registered with EPA. EPA intends that such certification would provide obligated parties with an "affirmative defense" if RINs are later found to be fraudulent—that is, obligated parties would not be liable for civil penalties under the Clean Air Act for the use of such RINs. As noted by Gina McCarthy, former EPA Assistant Administrator for Air and Radiation (now EPA Adminstrator), "the affirmative defense would ensure that refiners and other program participants who meet the conditions of the affirmative defense will not face civil penalties."[17]

A key component of a Quality Assurance Program would be the establishment of a quality assurance plan (QAP). The QAP would serve as the basis for audits by third party verifiers certified by EPA. EPA issued a proposed rule to create a QAP in January 2013 and projects the rule will be finalized in October 2013.[18]

Key questions include whether such an affirmative defense would also eliminate the requirement to purchase make-up RINs. While refiners and others would prefer to not "pay twice" for RINs, in general biofuel producers argue that not making up the RINs would undermine the legitimate RIN market. Simply put, if obligated parties are not required to replace invalid RINs with valid RINs, the size of the legitimate renewable fuel market is reduced. In EPA's proposed rule, nearly all invalid RINs would be required to be replaced by either the generator of the invalid RIN, the third party verifier, or the obligated party.

Policy Options

There are various policy options to address the issues of RIN fraud. EPA could undertake some of these options under existing Clean Air Act authority, while others would require congressional action. In general terms, there are at least four options:

1. Do nothing, and let market participants determine the credibility of actors they trade with;

2. Establish a Quality Assurance Program or some other certification to provide greater credibility, but do not tie it to EPA's determination on RIN validity;
3. Establish a certification procedure with an affirmative defense such that purchasers of invalid RINs are not liable for civil penalties (EPA's proposal); and
4. Establish a system where all certified RINs are valid for RFS compliance regardless of subsequent determination that they are fraudulent or otherwise deficient.

Currently, RIN market participants are acting under the first and second options. They are independently determining whether to trust the validity of the RINs they purchase. In many cases, after the RIN fraud came to light, obligated parties decided to purchase biodiesel and biodiesel RINs only from the largest producers.[19] At the same time, small producers have complained that they are unable to afford the verification procedures that some obligated parties now require.

Others have argued that the RIN fraud prosecutions have improved the integrity of the market. For example, one witness to a House Energy and Commerce Committee hearing on RIN fraud stated "[i]n essence, the Wild West of buying and selling RINs from market participants you don't know has ended, the wrongdoers are being rooted out, and everyone now knows that deals that are too good to be true are in fact too good to be true."[20]

Because of substantial remaining uncertainty about the integrity of the market, the National Biodiesel Board (NBB) and others are working to establish a more formal process for RIN certification. With the NBB, biodiesel producers have established a RIN Integrity Network where obligated parties can subscribe to a service where they can receive information on participating biodiesel producers.[21] This and similar efforts are in their early stages, and it is unclear how much credence RIN purchasers will give these networks. Further, it is unclear whether they will become the industry standard.

EPA is currently pursuing the third option, where a Quality Assurance Program is established with third parties auditing RIN generators. EPA has proposed an affirmative defense such that users of certified RINs would not be subject to civil penalties under the Clean Air Act. In most cases, under the proposal, the original RIN generator, the third party verifier, or the obligated party would be required to purchase additional RINs to make up for any certified RINs later found to be fraudulent. Obligated parties generally would prefer not to "pay twice" for RINs, adding to their compliance costs. However, eliminating this requirement would effectively shrink the market for biofuels under the RFS, harming legitimate biofuel producers. As EISA establishes specific fuel volume requirements, it is unclear whether EPA has the authority under existing statute to waive that requirement.

In addition to agency action, at least one bill has was introduced in the 112th Congress to amend the RIN system. H.R. 6444 would have required EPA to establish a RIN certification system by January 1, 2013. The bill would have precluded the agency from later invalidating any certified RINs. Thus, under the bill, any RIN found subsequently to be fraudulent would have still counted toward an obligated party's compliance, without penalties. As noted above, refiners and other obligated parties would have likely preferred this to other policy options, while biofuel producers were unlikely to support such blanket protection. No similar bills have been introduced in the 113th Congress.

Additional Questions

The concerns raised above, and proposed policy remedies, raise additional questions about the potential for RIN fraud in the future, as well as the effects on stakeholders from any policy solution. These questions include:

1. Beyond the instances of fraud currently being prosecuted, what other instances are unreported, and what other types of fraud are possible in the future?
2. How likely is fraud in the future, and what are the implications? and
3. How do various players benefit from the different policy options?

What Other Types of RIN Fraud Are Possible?

In the three cases of RIN fraud currently reported by EPA, fraudulent RINs were generated for fuel that did not exist. However, other potential errors or opportunities for fraud include:

1. double counting, where the same RIN (representing a gallon of actual fuel) is transferred to two different entities;
2. improper split, where a batch of RINs is separated into two or more groups and sold to different entities, but the total number of RINs somehow changes (an example of double counting);
3. improper reporting of RIN data (type of fuel, size of batch, etc.);

4. failure to report export (for every gallon of fuel exported any RINs from that export must be retired, as the fuel was never used as transportation fuel in the United States).

The EMTS was established in part to address errors of the first three types. However, it is unclear whether the EMTS completely screens out these errors. Especially as regulations require all transactions be reported within five business days, the real-time reporting requirements may potentially lead to errors if entities feel rushed in completing reports and transactions on time. On the other hand, real-time reporting may make it easier to catch errors and irregularities than under the previous system, where most data verification was completed on a quarterly basis.

The latter issue, that fuel has been exported without retirement of necessary RINs, has been raised by some stakeholders.[22] RFS regulations are explicit that when renewable fuel is exported that the exporter must have RINs to offset that volume: "Any party that owns any amount of renewable fuel, whether in its neat form or blended with gasoline or diesel, that is exported from any of the regions described in § 80.1426(b) shall acquire sufficient RINs to comply with all applicable Renewable Volume Obligations under paragraphs (b) through (e) of this section representing the exported renewable fuel."[23]

It is unclear to what extent, or whether, parties have been exporting fuel without securing the necessary RINs. EPA has not reported any such activity to date. To the extent that this sort of fraud is occurring, as with other types of fraud, it would lead to lower domestic renewable fuel use than required under the act. That would likely lead to an oversupply of RINs and a lower RIN price received by all market participants. However, unlike other types of fraud, actual fuel would be produced, so the overall level of U.S. biofuel production may not decline.

How Likely Is RIN Fraud in the Future?

To date, all of the reported cases of fraud have occurred in the biodiesel market. There are several reasons that have been given for this: (1) the market price for BBD RINs had been much higher than that for ethanol RINs, making any transaction (legal or illegal) that much more valuable; (2) in general biodiesel producers are smaller operations than ethanol producers, and the companies involved may be less well known to market participants; and (3) limited verification procedures existed at the time.

Between actions taken by industry, proposed regulations from EPA, and potential congressional action, the latter two reasons for fraud in the biodiesel RIN market may be fully addressed. However, the first cause—a high price for some RINs—may continue in the future. By 2022, the RFS requires the use of 36 billion gallons of renewable fuels, more than double the amount required in 2012. Thus, the absolute volume of the market will be larger in the future. Further, of the 36 billion gallons required in 2022, 16 billion gallons are required to come from cellulosic biofuels. Currently, there is very limited production of these fuels, and their production costs are high. To the extent that cellulosic fuel costs remain high in the future, the aggregate value of the cellulosic RIN market could be significantly higher than the total RIN market today. The higher value of this market might be a draw to actors looking to circumvent the law.

How Do Various Players Benefit from the Different Policy Options?

As noted above, obligated parties would like any new certification system to include an affirmative defense against civil penalties if they act in good faith.The details of that affirmative defense will determine who might benefit from policy changes. For example, in general obligated parties are more likely to benefit from a blanket exemption from both civil penalties and the requirement to purchase RINs to make up for ones later found to be invalid.[24] In general, biofuel producers are more likely to benefit from a policy that still requires obligated parties to purchase make-up RINs. Otherwise, the market for RINs—and thus the fuel they represent—would effectively shrink by the amount of any fraudulent RINs.

Any third-party certification procedures will add to the cost of producing biofuels and RINs. Various actors may be more or less able to absorb those costs. For example, obligated parties may be able to pass the additional cost along to gasoline and diesel fuel consumers through higher pump prices. Further, larger biofuel producers may be able to take advantage of economies of scale and spread the cost across all gallons of fuel they produce. Smaller producers, however, may be less able to spread the cost over the fewer gallons they produce and thus their per-gallon production costs may increase relative to their larger competitors. As noted above, many biodiesel producers are smaller operations who may feel more of these effects than ethanol producers (who generally produce larger volumes of fuel). In the future, cellulosic biofuel producers may also face similar pressures as these facilities are expected to be on the smaller side.

CONCLUSION

The establishment of the Renewable Fuel Standard has created a market for RINs that has grown both in volume and in value over time and is expected to continue to grow over the next decade. However, cases of fraud in the early years of the biodiesel RIN market raise questions about the integrity of RIN markets, as well as EPA's oversight of the markets. A dramatic increase in corn ethanol RINs in 2013 has raised additional questions about the RIN system and the overall RFS. Various policies have been proposed to address these concerns, and the details of those policies will affect RIN markets as well as the relative benefits to different market players.

End Notes

[1] For more information on the RFS, see CRS Report R40155, Renewable Fuel Standard (RFS): Overview and Issues, by Randy Schnepf and Brent D. Yacobucci.

[2] See CRS Report R41106, Meeting the Renewable Fuel Standard (RFS) Mandate for Cellulosic Biofuels: Questions and Answers, by Kelsi Bracmort.

[3] The initial RFS for cellulosic biofuels for 2012 was 500 million gallons. In December 2011 EPA decreased the mandate to 10.45 million ethanol-equivalent gallons. In January 2013, the U.S. Court of Appeals for D.C. vacated EPA's initial cellulosic mandate for 2012 and remanded EPA to replace it with a revised mandate. On February 28, 2013, EPA dropped the 2012 RFS for cellulosic biofuels to zero. For more information, see CRS Report R41106, Meeting the Renewable Fuel Standard (RFS) Mandate for Cellulosic Biofuels: Questions and Answers, by Kelsi Bracmort.

[4] The limitation on corn starch ethanol is roughly 8% on gasoline and diesel fuel combined.

[5] Thus, the effective cap on corn-based ethanol is 13.2 billion gallons in 2012, based on the difference between the overall mandate (15.2 billion gallons) and the advanced biofuel mandate (2.0 billion gallons).

[6] For more discussion on RINs see Robert Wisner, "Renewable Identification Numbers (RINs) and Government Biofuels Blending Mandates," AgMRC Renewable Energy Newsletter, Agricultural Marketing Research Center, Iowa State University, April 2009, available at http://www.agmrc.org/renewable_energy/ agmrc_renewable_energy_newsletter.cfm; or Wyatt Thompson, Seth Meyer, and Pat Westhoff, "Renewable Identification Numbers are the Tracking Instrument and Bellwether of U.S. Biofuel Mandates," EuroChoices 8(3), 2009, pp. 43-50.

[7] Unlike the other biofuel categories, the BBD mandate is a requirement on actual gallons. Thus, the 1.0 billion (actual) gallons required for the 2012 BBD mandate will generate 1.0 billion BBD RINs, but 1.5 billion advanced biofuel/renewable fuel RINs.

[8] In many cases, the RINs are detached from the actual fuel at the point of initial sale or transfer, and thus RINs may be detached for fuel that has not yet been blended into motor fuel or sold as motor fuel.

[9] EPA, "Regulation of Fuels and Fuel Additives: Changes to Renewable Fuel Standard Program; Final Rule," 75 Federal Register 14732, March 26, 2010.

[10] Ibid., p. 14733.

[11] Ryan Tracy, "U.S. Ethanol Mandate Puts Squeeze on Oil Refiners," The Wall Street Journal, March 10, 2013. For more information on the blend wall, see CRS Report R40445, Intermediate-Level Blends of Ethanol in Gasoline, and the Ethanol "Blend Wall", by Brent D. Yacobucci.

[12] Renewable Fuels Association, What Role Did Speculation Play in the March 2013 RIN Price Spike?, 2013.

[13] For example, OPIS reports daily RIN spot prices for four types of RINs, but does not report trading volume.

[14] NOVs available at EPA, Civil Enforcement of the Renewable Fuel Standard Program, http://www2.epa.gov/ enforcement/civil-enforcement-renewable-fuel-standard-program.

[15] On June 25, 2012, Rodney R. Hailey of Clean Green Fuels, LLC was found guilty of 8 counts of wire fraud, 32 counts of money laundering, and 2 counts of violating the Clean Air Act. On March 29, 2013, Jeffrey David Gunselman of Absolute Fuels, LLC pled guilty to 51 counts of wire fraud, 24 counts of money laundering and 4 counts of violating the Clean Air Act.

[16] 40 C.F.R. §80.1431(b)(2).

[17] Letter from Gina McCarthy, Assistant Administrator for Air and Radiation, EPA, to The Honorable Gene Green, Ranking Member, Ranking Member Subcommittee on Energy and Economy, Committee on Energy and Commerce, August 14, 2012.

[18] EPA, "RFS Renewable Identification Number (RIN) Quality Assurance Program; Proposed Rule," 78 Federal Register 12158, February 21, 2013.

[19] Testimony of various biofuel companies before the House Committee on Energy and Commerce Hearing on "RIN Fraud: EPA's Efforts to Ensure Market Integrity in the Renewable Fuels Program." July 11, 2012.

[20] Joe Jobe, Chief Executive Officer, National Biodiesel Board, Testimony Before the United States House of Representatives Committee on Energy and Commerce Hearing "RIN Fraud: EPA's Efforts to Ensure Market Integrity in the Renewable Fuels Program," Washington, DC, July 11, 2012.

[21] Genscape, Genscape's RIN Integrity Network, http://info.genscape.com/RIN.

[22] For example, see Jon P Fjeld-Hansen, Managing Director, Musket Corporation, Testimony Before the United States House of Representatives Committee on Energy and Commerce Hearing "RIN Fraud: EPA's Efforts to Ensure Market Integrity in the Renewable Fuels Program," Washington, DC, July 11, 2012.

[23] 40 C.F.R. §1430(a).

[24] H.R. 6444 (112th Congress) would have addressed this by removing EPA's authority to invalidate (for any reason) a certified RIN later found to be deficient. Thus, even a deficient RIN could be traded or used to meet a party's RVO as long as it had been certified by the agency.

In: The Renewable Fuel Standard
Editor: Lydia Rider
ISBN: 978-1-63321-829-1

Chapter 4

MEETING THE RENEWABLE FUEL STANDARD (RFS) MANDATE FOR CELLULOSIC BIOFUELS: QUESTIONS AND ANSWERS*

Kelsi Bracmort

SUMMARY

The Renewable Fuel Standard (RFS) was expanded under the Energy Independence and Security Act of 2007 (EISA; P.L. 110-140) in an effort to reduce dependence on foreign oil, promote biofuel use, and stabilize transportation fuel prices, among other goals. Over 15 years, the RFS requires that increasing amounts of biofuels—36 billion gallons by 2022—be used in transportation fuel. The mandate is to be accomplished in part with advanced biofuels, including cellulosic biofuels—fuels produced from cellulosic materials including grasses, trees, and agricultural and municipal wastes—which would increase over time to comprise some 44% of the RFS in 2022.

The U.S. Environmental Protection Agency (EPA) is required to set the annual standard for cellulosic biofuels under the RFS for the following year by November 30. If projected cellulosic biofuel production is less than the volume specified in the statute, EPA can lower

* This is an edited, reformatted and augmented version of a Congressional Research Service publication R41106, prepared for Members and Committees of Congress, dated December 9, 2013.

the cellulosic biofuels standard. EPA concluded that the nation lacked sufficient production capacity to meet the RFS cellulosic biofuels mandate for 2010, 2011, 2012, 2013, and 2014. EPA reduced the mandate for 2010 (from 100 million gallons to 5 million gallons actual volume), 2011 (from 250 million gallons to 6.6 million gallons), 2012 (from 500 million gallons to 8.65 million gallons, later vacated by a federal court decision and reduced to zero), and 2013 (from 1 billion gallons to 4 million gallons). EPA proposes to lower the 2014 mandate from 1.75 billion gallons to 17 million ethanol-equivalent gallons, and to rescind the 2011 cellulosic biofuel standard.

The 2010-2012 reduced mandates were not met by actual cellulosic biofuel production, which EPA reports was limited. Instead, these mandates were largely met with waiver credits. EPA reports that the cellulosic biofuels industry is growing incrementally, noting that two commercial-scale cellulosic biofuel facilities began fuel production in 2013, although it is unlikely that enough fuel will be supplied to meet the mandate.

The cellulosic biofuels industry may be able to produce enough fuel to meet the RFS mandates if certain obstacles are overcome: lowering the cost of conversion technology at the initial stages of commercial application, easing access to financing, removing feedstock supply uncertainties, and creating certainty for tax incentives. Another challenge for the cellulosic biofuels industry—and all biofuels industries—is the petroleum industry's opposition to the RFS overall, in part because it views the RFS as unworkable. Other industries—livestock and poultry producers in particular—have joined the petroleum industry in requesting that the RFS be modified, in many cases for reasons unrelated to cellulosic biofuel supply. Another supply constraint is the blend wall—the upper limit to the total amount of ethanol that by law can be blended into U.S. gasoline.

Several federal programs assist the cellulosic biofuels industry, including the U.S. Department of Agriculture's (USDA's) Biorefinery Assistance Program and Biomass Crop Assistance Program, and the U.S. Department of Energy's (DOE's) Loan Guarantee Program. EPA reports that some of the cellulosic biofuel companies identified in its 2014 proposed rule received or were offered significant federal financial support (approximately $387 million) in the form of grants and loan guarantees from USDA and DOE.

Many questions about cellulosic biofuels and the RFS have arisen. Can the RFS mandate for cellulosic biofuels be met? If so, when would it be met? What impact will the continued lowering of the cellulosic biofuels mandate by EPA have on investment in production? Should Congress continue to provide support for cellulosic biofuels, and if so, how? Might Congress statutorily increase the number of qualified feedstocks for the RFS cellulosic biofuels category, given the 112th

Congress amendment of the definition of cellulosic biofuels to include algae for some tax incentives? What impact will other legislative discussions (e.g., military support for biofuels) have on the RFS cellulosic biofuels mandate?

This report, in a question and answer format, discusses some challenges facing the cellulosic biofuels community, including feedstock supply estimates, and potential legislative options to address cellulosic biofuels production uncertainty for the RFS.

INTRODUCTION

The Renewable Fuel Standard (RFS), created by the Energy Policy Act of 2005 (EPAct05, P.L. 109-58) and expanded under the Energy Independence and Security Act of 2007 (EISA; P.L. 110- 140), mandates that domestic transportation fuel contain specified volumes of biofuels— including advanced biofuel, cellulosic biofuel, and biomass-based diesel—in an effort to reduce dependence on foreign oil, promote biofuel use, and stabilize transportation fuel prices, among other goals.[1] The mandate is to be accomplished in large part with cellulosic biofuels. The RFS requires that increasing amounts of biofuels be included in transportation fuel over a 15-year period, with the goal of using 36 billion gallons of biofuels annually by 2022, including 16 billion gallons of cellulosic biofuels.

The U.S. Environmental Protection Agency (EPA) is required to set the annual standard for cellulosic biofuels under the RFS for the following year by November 30. If the projected volume of cellulosic biofuel production is less than the volume specified in the statute, then EPA is required to lower the cellulosic biofuels standard. Due to a lack of U.S. production capacity, the EPA, in successive years from 2010 to 2013, issued final rules under its waiver authority for the RFS established in EISA, which lowered the 2010 cellulosic biofuel mandate from 100 million gallons to 5 million gallons (actual volume),[2] the 2011 mandate from 250 million gallons to 6.6 million gallons,[3] the 2012 mandate from 500 million gallons to 8.65 million gallons[4] (which was vacated by a federal court decision[5] and reduced to zero), and the 2013 mandate from 1 billion gallons to 4 million gallons.[6] EPA proposes to lower the 2014 mandate from 1.75 billion gallons to 17 million ethanol-equivalent gallons.[7] EPA also proposes to rescind the 2011 cellulosic biofuel standard.

Some have questioned whether cellulosic biofuels can be produced in sufficient quantity to satisfy the RFS mandate in future years. Feedstock supply, financial viability and financing, and technology advancement are considered among the limiting factors for cellulosic biofuel production. More than five years after EISA was enacted, progress toward meeting the cellulosic biofuels mandate has been delayed on multiple fronts (e.g., financial, administrative, and technical). Following four consecutive years of RFS cellulosic biofuels mandate reductions, some contend that Congress should reconsider the configuration of the RFS, determine whether additional resources are necessary for cellulosic biofuel production, and assess the success rate of this effort compared to other renewable energy efforts. Others assert that the current structure provides adequate incentives and measures to address any cellulosic biofuel productions shortfalls.

WHAT ARE CELLULOSIC BIOFUELS?

Cellulosic biofuels are liquid, solid, or gaseous fuels made from materials containing cellulose. Cellulose—a complex carbohydrate—is the organic matter found in plant walls that, along with hemicellulose and lignin, helps to give a plant its rigid structure. Cellulose feedstock includes agricultural residues (e.g., corn stover), forestry residues (e.g., wood chips), dedicated energy crops (e.g., switchgrass, hybrid poplar), and urban sources of waste (e.g., municipal solid waste).

The most widely discussed cellulosic biofuel is cellulosic ethanol for transportation.[8] Cellulosic ethanol differs from the corn-starch ethanol which dominates transportation biofuels in the United States; it is made from feedstock with no or only limited food value, potentially results in fewer greenhouse gas emissions, and has a higher lifecycle energy balance.[9] Converting cellulosic feedstock to ethanol, however, is more expensive and difficult than converting corn starch to ethanol. The conversion of cellulose to ethanol generally happens in three phases—pretreatment, hydrolysis, and fermentation to ethanol. Pretreatment weakens the plant wall structure so that the cellulose is easier to obtain during hydrolysis. Hydrolysis—acid or enzymatic—separates the cellulose into sugars. Fermentation converts the sugars into ethanol. Cellulose can also be converted to liquid fuels through processes other than fermentation (e.g., thermochemical processes).[10]

Proponents suggest that increased use of cellulosic biofuels for transportation could potentially help to reduce U.S. dependence on foreign oil, strengthen rural economies, and improve the environment. In addition, they contend that cellulosic feedstocks may fare better in the food-energy debate, since either non-food crops, crop residues, or other wastes are used for cellulosic biofuel production, instead of food or feed commodities. In contrast, others argue that cellulosic biofuels at RFS volumes require a substantial feedstock supply that has yet to be verified and may never be cost-competitive without government support.

WHAT IS THE RELATIONSHIP BETWEEN CELLULOSIC BIOFUELS AND THE RENEWABLE FUEL STANDARD?

The RFS established in Section 202 of EISA called for 100 million gallons of cellulosic biofuels to be included in the national transportation fuel supply in 2010 (see *Table 1*). The mandate increases to 16 billion gallons by 2022.[11] Despite the RFS mandate for cellulosic biofuels starting in 2010, the first registered production of cellulosic biofuels in the United States was not achieved until 2012.[12]

Table 1. RFS Biofuels Mandate in EISA: Total Renewable Fuels and Cellulosic Biofuels (in billions of gallons)

		Cellulosic biofuels	
Year	Total renewable fuels	Original	EPA revised (actual volume)
2006	—	—	—
2007	—	—	—
2008	9.00	0.00	—
2009	11.10	0.00	—
2010	12.95	0.100	0.005[a]
2011	13.95	0.250	0.0066[b]
2012	15.20	0.500	0[c]
2013	16.55	1.00	0.004[d]
2014	18.15 (15.21 EPA revised)[e]	1.75	<0.010[e]

Source: EISA (P.L. 110-140), Section 202.

[a] U.S. Environmental Protection Agency, "Regulation of Fuels and Fuel Additives: Changes to Renewable Fuel Standard Program; Final Rule," 75 *Federal Register*, March 26, 2010.

[b] U.S. Environmental Protection Agency, "Regulation of Fuels and Fuel Additives: 2011 Renewable Fuel Standards; Final Rule," *Federal Register*, December 9, 2010.

[c] Vacated under *API vs. EPA*. Level originally set by EPA at 0.00865 billion gallons. U.S. Environmental Protection Agency, "Regulation of Fuels and Fuel Additives: 2012 Renewable Fuel Standards; Final Rule," *Federal Register*, January 9, 2012.

[d] U.S. Environmental Protection Agency, "Regulation of Fuels and Fuel Additives: 2013 Renewable Fuel Standards; Final Rule," 78 *Federal Register*, August 15, 2013.

[e] EPA proposes to reduce the 2014 cellulosic biofuel mandate to 17 million ethanol-equivalent gallons. EPA does not state the actual volume amount in the proposed rule. CRS analysis suggests that the mandate's actual volume will be less than 10 million gallons. U.S. Environmental Protection Agency, U.S. Environmental Protection Agency, "2014 Standards for the Renewable Fuel Standard Program; Proposed Rule," 78 *Federal Register*, November 29, 2013.

Data and analysis presented during the RFS debate and ultimate passage of EISA in 2007 supported the idea that these levels of cellulosic biofuel production capacity could be achievable. Some argued that plentiful feedstock would be available[13] and that the conversion technology was close to being certified as commercially viable. Moreover, some presumed that the federal government would provide substantial financial support and enhance the infrastructure needed to spur a commercial cellulosic biofuels market.[14] Others were leery about the time frame provided to meet the RFS cellulosic biofuels mandate given the as-yet-undeveloped production capacity.[15]

EPA has the authority to waive completely or in part the consumption mandates for cellulosic biofuels established in EISA, under certain circumstances.[16] For instance, the cellulosic biofuels requirement may be waived if the Administrator determines, after public notice and opportunity for comment, that there is an inadequate domestic supply.[17] EPA may also waive the RFS or any specific category of the RFS if there is evidence that full implementation may cause severe economic or environmental harm. EPA generally conducts its cellulosic biofuels volume analysis based on an evaluation of individual producers' production plans and progress to date, following discussions with cellulosic biofuel producers, the Energy Information Administration (EIA), the Department of Agriculture (USDA), and the Department of Energy, and an assessment of the probabilities associated with production schedules from each of these producers.[18]

What Challenges Are Associated with Production of Cellulosic Biofuels?

Since the RFS was expanded under EISA,[19] U.S. production of cellulosic biofuels has had a slow start. Impediments to increasing capacity include technology setbacks and a lack of financial support.

Commercial cellulosic biofuel facilities are estimated to cost hundreds of millions of dollars, significantly more than a traditional corn ethanol plant, especially when measured in terms of the amount of biofuel produced (measured in million gallons per year, or mgy). For example, the company INEOS Bio reports that its Indian River 8 mgy production capacity plant cost approximately $130 million.[20] American Process, Inc., reports that its Alpena 0.8 mgy pilot plant cost approximately $36 million.[21] In comparison, a 40 mgy corn ethanol plant cost approximately $80 million to construct in 2006.[22] Some lenders find it extremely risky, perhaps even cost-prohibitive, to provide financial backing to cellulosic biofuel plants, mainly because the conversion technology has not been applied or proven on a commercial scale.[23] These costs, and the perceived financial risk, are likely to decrease if a significant number of commercial-scale plants come online and prove to be economically viable.

In addition to financing issues, other challenges to cellulosic biofuel production include multiple definitions of biomass in various laws, which may confuse investors and lead to legal disputes over what constitutes a "qualified feedstock." For example, incentives for cellulosic biofuel production in the tax code are relatively broad. The renewable biomass definition for the RFS under EISA does not allow for biomass removed from federal lands, and excludes crops from forested lands.[24] Some argue that opening federal lands for biomass removal could provide an inexpensive supply of cellulosic feedstock that would be immediately available to biorefineries for cellulosic biofuel production. Others contend that biomass removal from federal lands is a short-term response to the cellulosic feedstock source problem and might not be carried out in a sustainable manner, leading to deterioration of the nation's parks and recreation areas. The definition of biomass under EISA also excludes most municipal solid waste (MSW), which some view as a potential source for conversion to biofuels.

Also challenging for cellulosic biofuel production are feedstock contracts, which will likely cover multiple years. Agricultural and forestry producers may not agree to a contract for a cellulosic feedstock that requires a lengthy time commitment. For example, it generally takes three years for switchgrass crops to reach maturity.[25] As a result, a producer would have to commit their land to one particular cellulosic feedstock crop for a number of years, thus limiting the producer's choice to grow other, potentially more remunerative crops during that period. The Biomass Crop Assistance Program (BCAP) is designed to assist producers with annual payments that could mitigate some of the risks involved with shifting from traditional crop production to energy crop production. In early 2013, EPA reported that the industry has made significant progress regarding feedstock supply, citing companies that have secured feedstock contracts with local agricultural producers.[26]

HOW MUCH CELLULOSIC FEEDSTOCK EXISTS FOR CONVERSION TO BIOFUELS?

An important factor in evaluating whether a commercial cellulosic biofuel production plant will be a favorable investment is whether a steady feedstock supply exists for a biorefinery.[27] However, determining actual availability of feedstock is difficult. Quantifying the amount of feedstock available for conversion to biofuels requires information about feedstock sources, production rates, accessibility, and location restrictions (e.g., public versus private lands if the feedstock is to be used for certain energy purposes). Investors must make feedstock predictions based on data from weather patterns and land use change, as well as an uncertain supply chain (e.g., handling, storage, and transportation costs), among other things. This is particularly important when a biomass feedstock growth season of four to five months must provide biomass feedstock for 12 months of plant operation.[28] To make the supply available throughout the year, special equipment may be required for feedstock harvest, handling, collection, storage, or transport, as cellulosic feedstock is often too bulky for standard farming equipment to handle. Some of the concerns described above may not be applicable to facilities that use certain types of wastes (e.g., municipal solid wastes) whose supply doesn't depend on as many variables and can be provided at low to no cost relative to other cellulosic feedstocks.

The amount of cellulosic feedstock necessary for conversion to a biofuel depends on the feedstock type, the conversion process, and the desired biofuel.[29] Biofuel conversion yield is measured in gallons per ton. For harvested crops (as opposed to wastes), feedstock, or crop, yield is measured in tons per acre. Total yield, measured in gallons per acre, depends on both the feedstock yield and the conversion yield. Yields are typically much higher on prime cropland than on marginal lands. However, if cellulosic feedstocks are produced on prime cropland, then they likely compete with food crops while significantly raising their cost of production due to the opportunity cost of land. Some argue that current estimated cellulosic feedstock yields will need to increase markedly over the next decade to meet the RFS mandate of 16 billion gallons of cellulosic biofuel production per year by 2022. Others contend that a significant growth of cellulosic feedstock is not essential, as advances in conversion technologies will afford the opportunity to produce more cellulosic biofuel with less feedstock.

WHAT FEDERAL FINANCIAL ASSISTANCE IS AVAILABLE FOR THE CELLULOSIC BIOFUELS INDUSTRY?

Several forms of federal government financial support exist for production of cellulosic biofuels, including the DOE Loan Guarantee Program, the USDA Biorefinery Assistance Program, the USDA Biomass Crop Assistance Program, and tax incentives.[30] Other sources of financial assistance include smaller grants awarded by federal agencies for unique initiatives.[31]

To help promote cleaner energy technologies, including cellulosic biofuel production technologies, Congress established the Department of Energy (DOE) Loan Guarantee Program (LGP).[32] Loans may not exceed 80% of total project costs. The DOE Section 1703 loan program supports clean energy technologies that are high technology risks (e.g., biomass, solar, alternative fuel vehicle technologies). In addition, the DOE Section 1705 loan program, which expired in 2011, supported renewable energy systems, electric power transmission systems, and leading-edge biofuels.[33] Some in the industry are concerned that the LGP is not being carried out at a pace responsive to market momentum for cellulosic biofuels.[34]

The Government Accountability Office (GAO) reviewed DOE's execution of the Loan Guarantee Program in 2010 and again in 2012.[35] In 2012, GAO found that DOE "improved the speed at which it was able to move section 1705 applications through its review process" and that DOE had "not closed any section 1703 loan guarantees or otherwise demonstrated that the program is fully functional." Even though Section 1705 expired in 2011, GAO issued three recommendations for the overall loan guarantee program, including that DOE "implement a consolidated system that enables the tracking of the status of applications and that measures overall program performance."

The USDA Biorefinery Assistance Program (BAP) assists in the development of new and emerging technologies for advanced biofuels.[36] BAP provides competitive grants and loan guarantees for construction and/or retrofitting of demonstration-scale biorefineries to demonstrate the commercial viability of one or more processes for converting renewable biomass to advanced biofuels. Because BAP has not received any discretionary funding needed to implement the grant portion of the program, USDA has only implemented the loan guarantee portion of the program using mandatory funding. USDA reports that seven biorefineries were granted a BAP conditional commitment for a loan guarantee at a total of approximately $704 million from FY2009 through FY2012.[37]

Another federal government program, focused on feedstock development, is the USDA Biomass Crop Assistance Program (BCAP).[38] The two main program objectives of BCAP are to support the establishment and production of eligible crops for conversion to bioenergy in selected areas, and to assist agricultural and forest land owners and operators with collection, harvest, storage, and transportation of eligible materials for use in a biomass conversion facility. USDA issued the BCAP final rule on October 27, 2010, implementing both program components. BCAP was one of several farm bill energy programs set to expire at the end of FY2012. The American Taxpayer Relief Act of 2012 (ATRA, P.L. 112-240) extended BCAP through FY2013.[39] Since the program's inception, 11 BCAP project areas have been established.[40] According to USDA, BCAP matching payments[41] totaled $557,443 for the transportation of nearly 10,320 dry tons of herbaceous crop residue (i.e., corn stover) in FY2012.[42]

Additionally, the government provides tax incentives to support cellulosic biofuel production.[43] The 2008 farm bill offers a production credit of $1.01 per gallon of biofuel produced from qualifying second-generation biofuel feedstocks, including cellulosic matter. The credit expired on December 31, 2012, but ATRA extended it and expanded the credit to include fuels produced from algae. The credit is set to expire at the end of calendar year 2013. There is also a special depreciation allowance for a new cellulosic biofuel plant that originated under the Tax Relief and Health Care Act of 2006 (P.L. 109-432). ATRA amended the allowance to include algae as a qualified feedstock for the special allowance. The allowance also expires at the end of 2013.

WERE THE REVISED 2010, 2011, AND 2012 RFS MANDATES FOR CELLULOSIC BIOFUELS MET? WILL THE 2013 MANDATE AND THE PROPOSED 2014 MANDATE BE MET?

EPA's analyses concluded that the United States did not have sufficient cellulosic biofuel production capacity to meet the scheduled RFS mandates for 2010-2014. Therefore, EPA reduced the mandate (actual volume)[44] for 2010 from 100 million gallons to 5 million gallons,[45] the mandate for 2011 from 250 million gallons to 6.6 million gallons,[46] the mandate for 2012 from 500 million gallons to 8.65 million gallons,[47] and the mandate for 2013 from 1 billion gallons to 4 million gallons.[48] EPA proposes to lower the 2014 mandate from 1.75 billion gallons to 17 million ethanol-equivalent gallons.[49] EPA also proposes to rescind the 2011 cellulosic biofuel standard, partially in response to the decision in *API vs. EPA*.

EPA monitors RFS compliance by fuel blenders according to their reported accumulation of renewable identification numbers (RINs) for each biofuel's yearly volumetric standard. A RIN is a unique 38-character number that is issued (in accordance with EPA guidelines) by the biofuel producer or importer at the point of biofuel production or the port of importation. Each qualifying gallon of renewable fuel has its own unique RIN. The EPA uses its Moderated Transaction System (EMTS) to manage RIN transactions.[50] Using data generated from EMTS, EPA provides aggregated monthly data on RIN generation and renewable fuel volume production for specific fuel categories, such as cellulosic biofuel.[51] In general, compliance is achieved by a fuel blender when the number of RINS generated and assigned each year for cellulosic biofuels equate to the volume requirement for that year.

As noted above, EPA has the authority to waive the cellulosic biofuel mandate on a yearly basis. In any year that EPA grants a waiver, the agency must also make waiver credits available at a set price, the formula for which is set in statute.[52] Waiver credits may be used by obligated parties to comply with the cellulosic biofuel volume obligation in lieu of RINs generated with the production of the cellulosic biofuel.[53] The 2010, 2011, 2012, and 2013 prices for a cellulosic biofuel waiver credit are $1.56, $1.13, $0.78, and $0.42, respectively. EPA did not announce the price for the 2014 cellulosic biofuel waiver credit in its proposed rule. If waiver credits[54] are issued (when the mandate is reduced), compliance is met when the combination of waiver credits and RINs equate to the volume requirement for that year. Previous-year

RINs may be used to meet current-year compliance, although there are some stipulations.

Even before enactment of EISA, which expanded the RFS, reported production data indicated that overcoming any or all of the hurdles to increasing cellulosic biofuel production to meet the original 2010 RFS mandate of 100 million gallons set by Congress was unlikely. 2010 was a transition year for the RFS, going from the RFS established under EPAct05 (RFS1) to the RFS established under EISA (RFS2). RFS1 had different production requirements than RFS2 for cellulosic biofuels,[55] including an ethanol-equivalency ratio of 2.5-to-1, where each gallon of cellulosic ethanol counted as 2.5 gallons toward the EPAct05 mandate.[56] EPA reports that zero cellulosic biofuel RINs were produced in 2010 under the RFS2 rules.[57] EPA reports that the majority of compliance for 2010 was met with RINs generated (i.e., production) under RFS1.[58] EMTS records show that 12,186 cellulosic biofuel waiver credits were purchased in 2010, which likely made up for the remainder of 2010 compliance.[59]

More compliance data are available for the 2011 and 2012 RFS cellulosic biofuels mandates. However, evaluation of the available data is difficult. Compliance for both years could have been met with cellulosic biofuel RINs and waiver credits. EPA reports that zero cellulosic biofuel RINS were produced in 2011.[60] Compliance for 2011 was met with excess cellulosic biofuel RINs generated in 2010 under RFS1 and cellulosic biofuel waiver credits.[61] EMTS records report that 4,248,388 RFS2 cellulosic biofuel waiver credits were purchased in 2011.[62] EPA reports that there is no "need to address compliance with the 2012 cellulosic biofuel standard" since there was a federal court decision that vacated the 2012 standard.[63] EMTS reports that 20,069 RINS were generated for cellulosic biofuel and 1,741 RINS were generated for cellulosic diesel in 2012.[64]

In its 2013 final rule, EPA set a 2013 cellulosic biofuel actual volume standard of 4 million gallons (actual volume) that is based on the Energy Information Administration's (EIA's) estimate of available supply,[65] production volumes developed in consultation with the companies expected to produce cellulosic biofuel from commercial-scale facilities in 2013, and EPA's judgment.[66] EPA projected that the 2013 cellulosic mandate would be met by production from two companies: INEOS Bio and KiOR. [67] Thus far, EMTS reports that 294,554 RINs were generated for cellulosic biofuel and 269,260 RINS were generated for cellulosic diesel through October 2013.

In its 2014 proposed rule, EPA proposes a 2014 cellulosic biofuel standard of 17 million ethanol-equivalent gallons. EPA projects that the mandate could be met by five companies with approved pathways, two of which are operational (INEOS Bio and KiOR). Additionally, EPA proposes that the mandate may be met by potential cellulosic biofuel producers with proposed pathways (e.g., compressed natural gas and liquefied natural gas from landfill biogas if used as a transportation fuel, cellulosic ethanol from corn kernel fiber, and renewable fuel oil if used as heating oil). Last, EPA acknowledges that there are potential foreign sources of cellulosic biofuels, although these entities are not expected to export the fuel they produce to the United States. EPA reports that there are four foreign companies with approved fuel production pathways, but the companies have not registered their cellulosic biofuel facilities with the RFS. EPA made similar assessments of some of these same plants in its 2013 rule—assessments that proved inaccurate. It is not evident what has changed between the time the 2013 rule was finalized and the 2014 rule was proposed.

What Impact Will Significantly Lowering the RFS Mandates Have on Investment in Cellulosic Biofuel Production?

EPA's waiver authority creates uncertainty for investors in cellulosic biofuel ventures. Investors may fear that the full cellulosic biofuels mandate will continually be waived to lower amounts by EPA, thus depriving them of the government-mandated market on which they had originally based their investment. EPA's 2014 proposal to lower the overall RFS may further hamper investment.

Those in favor of continuing an RFS with a cellulosic biofuels component argue that adequate production and consumption of cellulosic biofuels can be achieved in a cost-effective manner if there is consistent federal policy, possibly allaying the concerns of some investors.[68] However, minimal actual production has led some to question the viability of cellulosic biofuels over the long term.[69] Furthermore, a cellulosic biofuels industry that has regularly missed the reduced mandates has drawn criticism from groups—especially the petroleum industry—that contend they are being unfairly targeted to purchase credits for a fuel that is unlikely to be produced at the required levels.[70]

Another form of uncertainty for all involved parties is the delayed announcement of the proposed annual renewable fuel standards. Under the Clean Air Act, each year's standards are required to be finalized by November

30 of the previous year, although the statute does not stipulate any penalty or other action if this deadline is missed. EPA was two months late in proposing, let alone finalizing, the 2013 standard. The delay may lead to uncertainty in investments and planning for the biofuel and petroleum industries, among others. In its 2013 final rule, EPA announced it will extend the compliance deadline for the 2013 RFS standards from February 28, 2014, to June 30, 2014, to "alleviate some of the concerns that obligated parties have regarding the tardiness of the final rule and its effect on their decisions regarding RIN acquisition."[71] EPA issued the proposed 2014 standards prior to the November 30 deadline, although with a 60-day comment period, the final rule may not be issued before late January 2014.

Investors and the cellulosic biofuels community alike may be more apprehensive about the federal policy for advanced biofuels as 2016 draws nearer. There is a provision in the RFS statute that requires the EPA Administrator, starting in 2016, to modify the applicable volumes of the RFS in its entirety for subsequent years if the Administrator waives the renewable fuel mandate, the advanced biofuel mandate, the cellulosic biofuels mandate, or the biomass-based diesel mandate by at least 20% for two consecutive years or by at least 50% for a single year.[72] As a result, in 2016, some biofuel industries (e.g., corn ethanol) that are satisfying their mandates may find that they are being penalized because of another biofuel industry (e.g., cellulosic biofuel) that is not meeting its specified mandate.[73]

HOW MANY COMMERCIAL CELLULOSIC BIOFUEL PLANTS EXIST?

EPA reports that cellulosic biofuel production in the United States remains limited.[74] In its 2014 proposed rule, EPA provides information on 11 companies and one additional industry (compressed and liquified natural gas) with the potential to produce cellulosic biofuels.[75] EPA's assessment includes their biofuel production estimates, technology process, and anticipated start dates. Four of the companies identified in the 2014 proposed rule were also identified in the 2013 final rule. EPA reports that two facilities commenced operation in 2013: KiOR and INEOS Bio. Biofuel production pathways have not been approved by EPA for one of the companies or for the compressed and liquified natural gas industry.

As observed over the last few years, due to unforeseen financial and technical issues, it can be difficult to gauge when or if cellulosic biofuel companies will actually come online and how much cellulosic biofuel will be produced. In 2010, it was reported that approximately two dozen demonstration- or pilot-scale cellulosic ethanol plants existed in the United States.[76] The Advanced Ethanol Council reported in 2012 that there were nine cellulosic biofuel pilot/demonstration facilities, eight commercial facilities under construction or commissioning, and seven commercial facilities at the engineering stage.[77] Environmental Entrepreneurs—a national community of business leaders—projects that 18 commercial cellulosic ethanol facilities will come online in the United States by 2016.[78]

WHAT'S NEXT FOR THE RFS CELLULOSIC BIOFUELS MANDATE?

The shift toward large-scale, economically feasible production of cellulosic biofuels has thus far proven to be illusive. Advancements have been made, but not in step with the pace set by Congress for the RFS. Multiple factors have contributed to the cellulosic biofuels market being at the stage it is today, with the two major factors being technological issues and a recession that slowed the economy, including energy demand. Due to production shortfalls, EPA has significantly reduced the cellulosic biofuels mandate each year from 2010 through 2013 and proposes to do so again for 2014. It appears that Congress overestimated how much cellulosic biofuel production would occur on an annual basis. It also appears that projecting the production amount continues to be a challenge for both EPA and EIA.[79] Thus, it is difficult to state with certainty how much cellulosic biofuel may be produced in the future and in what time frame.

If mandate reductions and continual production shortfalls signify that the cellulosic biofuels mandate is not working as intended, Congress could decide modifications are necessary. Advocates of cellulosic biofuels may find it beneficial for Congress to make the mandate more attainable in the near term (e.g., two to five years), primarily so that the industry has more time to show whether large-scale commercial production is economically feasible. Opponents may view any additional congressional action to assist the cellulosic biofuels community as harmful to the entire renewable energy market in the long run, as some believe that it could be extending the life of dead-end technologies.

The Role of Congress

Congress will continue to be the final arbiter over the future of the RFS. Should the RFS be continued in its present form, abridged, or eliminated completely? What additional public costs would each of these options engender? The discussion about whether the cellulosic biofuels industry requires additional federal support to meet the RFS mandate has occurred since the inception of the RFS; however, the discussion became more pronounced when the RFS was expanded in 2007 to call for a much larger volume of cellulosic biofuel to be produced in a shorter timeframe. In response, Members of the 111th, 112th, and 113th Congresses have proposed a variety of bills that could affect the cellulosic biofuels industry (see *Table 2*). The 112th Congress passed legislation that may aid the industry. In particular, the American Taxpayer Relief Act of 2012 (ATRA, P.L. 112-240) extended the cellulosic biofuel production tax credit and the cellulosic biofuel plant special depreciation allowance through the end of 2013, and included algae as a qualified feedstock for both tax incentives. ATRA also renamed the tax credit and depreciation allowance to "second generation" biofuel, removing the term "cellulosic."

Table 2. Selected Proposed Legislation from the 113th, 112th, and 111th Congresses Pertaining to Cellulosic Biofuels

Congress		Legislation
113th	•	H.R. 550 and S. 251 would modify the way EPA projects yearly cellulosic biofuel mandates.
	•	H.R. 796 would require a reduction in the advanced biofuel
		mandate if the cellulosic mandate is lowered in any given year.
	•	H.R. 1462 would eliminate the non-advanced biofuel portion of the
		RFS and would modify the way EPA projects yearly cellulosic biofuel mandates.
112th	•	H.R. 230 would have modified the DOE Loan Guarantee Program, appropriating funds for EPAct05 Section 1512 Conversion Assistance for Cellulosic Biomass Waste-Derived Ethanol
		Approved Renewable Fuels, establishing a loan guarantee program for cellulosic ethanol production technology development.

Congress		Legislation
	•	H.R. 424 would have amended the RFS to revert back to RFS1 targets and eliminated all advanced biofuel mandates.
	•	H.R. 851, H.R. 2231, S. 884, S. 1294 would have implemented new financial support mechanisms, extending relevant tax provisions.
	•	H.R. 6047 would have modified the way EPA projects yearly cellulosic biofuel volume mandates.
111th	•	S. 943 and H.R. 2283 would have waived the lifecycle greenhouse gas emission reduction requirements for renewable fuel production.
	•	Section 129 of the American Clean Energy and Security Act of 2009 (H.R. 2454, also known as Waxman-Markey) would have amended the Loan Guarantee Program to incorporate renewable fuel pipeline construction.
	•	Title I of the American Clean Energy Leadership Act of 2009 (ACELA, S. 1462) would have amended the Loan Guarantee Program and created a Clean Energy Deployment Administration, under DOE, to advance lending and implementation of commercial clean energy technologies.

Source: CRS.
Notes: Legislation is presented in order of House introduction.

One option for Congress could be modification of the cellulosic biofuels mandate, and thus the RFS. This option could require an analysis of the cellulosic biofuels market and, perhaps more importantly, consideration of circumstances that were not as prevalent when the RFS was expanded in 2007. Some of these circumstances include budget concerns, the blend wall,[80] military interest in renewable fuels,[81] and the slow development of the cellulosic biofuels production sector. Such an analysis might quantify the amount of federal support already granted to the cellulosic biofuels industry. EPA reports that four of the ten production facilities with approved cellulosic biofuel pathways have received or were offered significant federal financial support (approximately $387 million) in the form of grants and loan guarantees from USDA and DOE.[82] One analysis of federal and public funding data for advanced biofuel projects provided by Environmental Entrepreneurs suggest that, since 2009, USDA DOE and other federal entities have provided or offered some $734 million in grants and loan guarantees to eight companies that Environmental Entrepreneurs expects to have cellulosic ethanol facilities come online by 2016.[83]

When expanding the RFS in EISA, Congress in effect made a technology choice by selecting a preferred technology (i.e., cellulosic biofuel) among many unknown technologies. Undoing that commitment could involve substantial costs to both public and private investors. Another issue to be analyzed may be a technology assessment to determine if selecting one unproven technology for financial and policy incentives fits into a larger energy policy context.

Alternatively, Congress could take a wait-and-see approach to addressing concerns with the cellulosic biofuels mandate. Given EPA's consecutive lowering of the mandate by approximately 95% to 99% for each of the last four years, some question whether the cellulosic biofuels industry can ramp up production over the next two years to meet the scheduled RFS mandates. There may be an upcoming shift where some corn ethanol facilities plan to transition to cellulosic biofuel production, and it is not known what effect this might have on cellulosic biofuel production estimates.[84] Further, some anticipate that additional feedstocks and production pathways may qualify in the near future for the RFS cellulosic biofuel mandate. For instance, in February 2013, EPA released a final rule that identifies additional qualified fuel pathways to generate cellulosic biofuels, and allows renewable gasoline and renewable gasoline blendstock made from certain feedstocks to qualify as cellulosic biofuels.[85]

Congress may find that a substantial portion of the cellulosic biofuels volume mandate is unattainable for years to come. EIA reports that while "cellulosic biofuels volumes are expected to grow significantly relative to current levels, they will likely remain well below the targets envisioned in the Energy Independence and Security Act of 2007."[86] Congress has multiple options to address the lack of cellulosic biofuels production. Congress could make a statutory change to the definition of cellulosic biofuels for the RFS to open it up to additional feedstocks that can assist with meeting the annual production targets. Or Congress may continue to require the EPA to set the cellulosic biofuels mandate based on the best available evidence—the status quo option.[87] In the 113th Congress, some Members have proposed legislation (e.g., H.R. 550, S. 251) that would require EPA to base the cellulosic mandate on actual production from the previous year. Congress could also eliminate the cellulosic biofuels portion of the mandate, or the RFS entirely.

The Role of EPA

Congress directed EPA to implement the RFS. After assessing the cellulosic biofuels market, consulting with EIA, and taking other measures, EPA has lowered the cellulosic biofuel mandate when the scheduled mandate could not be met. Some contend that EPA has not carried out its responsibilities regarding the RFS cellulosic biofuels standard as expeditiously or accurately as desired. Some are not satisfied with the success rate of meeting the cellulosic biofuels standard. Some could argue that EPA has a difficult job trying to follow through on Congress's directions, given the challenges in the cellulosic biofuels industry. Further, it could be argued that any federal agency would likely find it difficult to accurately project when a new technology selected by Congress, not yet proven at commercial scale, would come online and at what pace. EPA can modify its projection procedure for cellulosic biofuels, but any major changes to its procedure would likely require additional instruction from Congress.

With regard to the 2012 standard, a federal court responded to the question of whether EPA is using an appropriate methodology to estimate the amount of cellulosic biofuels that can be produced. The court ruling contains principles that EPA will have to apply in future standards as well. EPA accepted the federal court's decision to vacate the 2012 cellulosic biofuel standard, and has proposed rescinding the 2011 standard as well.[88] The court discussed, at length, one significant factor that EPA, in its 2013 final rule, appears to address—that is, EPA's responsibility to accurately project how much cellulosic biofuel will be produced, not to "individually advance a technology-forcing [cellulosic biofuels] agenda."[89]

There are at least three significant distinctions in the 2013 final rule compared to previous rules: the changes made in projecting cellulosic biofuel production, the discussions regarding feedstock supply, and the facility start dates and corresponding ramp-up periods. EPA reports it made several changes to the way the agency projects cellulosic biofuel production based on the recent federal court ruling and to help "ensure a neutral projection that aims at accuracy." Some of the projection changes were not including production volumes from pilot and demonstration scale facilities, not including production volumes from facilities with start-up dates near the end of the year, and setting benchmarks for how quickly EPA assumes new facilities will ramp up to full production. In its 2013 final rule, EPA discussed feedstock supply more thoroughly than in previous rules. EPA noted that feedstock agreements or contracts have been reached for structurally complete facilities and for facilities expected to be complete by the end of the year.[90] Additionally, EPA noted the possibility for cellulosic biofuel to be produced from other sources, such as from corn fiber or landfill biogas, depending on if and when these pathways are approved by EPA. However, it appears that, despite these changes, EPA has again overestimated the cellulosic biofuel production level for 2013.

The 2014 proposed rule builds on the information given in the 2013 final rule, and gives more insight on certain factors that impact the 2014 cellulosic biofuel volume projection and how that projection may be met. For example, there is a new section with details about modeling the uncertainty associated with estimated cellulosic biofuel production ranges from numerous facilities. EPA explains how it proposes to take into account uncertainty due to the completion and commissioning phases of a facility, the ramp-up phase of new facilities, and the ability of a facility to maintain consistent production. Further, EPA explores how potentially the 2014 mandate may be met if a new pathway—the production of compressed or liquified natural gas from landfill biogas if used as transportation fuel—is approved.

EPA has not been given much guidance by the court, or Congress, on how to calculate a cellulosic biofuels projection, although the methodology the agency used previously was under scrutiny from the oil industry, among others, as evidenced by the recent court case concerning the 2012 cellulosic biofuels production projection and a pending court case addressing the 2011 projection. It is not clear if the current methodology being used for the 2014 projections will face the same scrutiny. If a projection is repeatedly not met and sources enlisted for assistance provide projections that turn out to be wrong (e.g., EIA, cellulosic biofuel industry), should EPA at some point have set the cellulosic biofuels standard at close to zero (e.g., 100,000 gallons) until actual commercial production of a certain threshold was reached? Arguably, setting a standard too low could impact federal support (e.g., grants, tax incentives) and private investment for the technology. This could have hindered the overall goals of the RFS, economic development in rural areas, and more. Prospects for the cellulosic biofuels industry could be affected if EPA continues to incorrectly project annual amounts of cellulosic biofuel production.

End Notes

[1] For more information on the expanded RFS, see CRS Report R43325, The Renewable Fuel Standard: In Brief; and CRS Report R40155, Renewable Fuel Standard (RFS): Overview and Issues.

[2] U.S. Environmental Protection Agency, "Regulation of Fuels and Fuel Additives: Changes to Renewable Fuel Standard Program; Final Rule," 75 Federal Register, March 26, 2010. Actual volume is the true physical volume of the fuel. Ethanol-equivalent volume is the

volume used for RFS compliance purposes, which takes into account the energy content of the fuel.

3 U.S. Environmental Protection Agency, "Regulation of Fuels and Fuel Additives: 2011 Renewable Fuel Standards; Final Rule," Federal Register, December 9, 2010.

4 U.S. Environmental Protection Agency, "Regulation of Fuels and Fuel Additives: 2012 Renewable Fuel Standards; Final Rule," Federal Register, January 9, 2012.

5 American Petroleum Institute v. EPA, 706 F.3d 474 (D.C. Cir. 2013); American Petroleum Institute v. Environmental Protection Agency, USCA Case #12-1139, January 25, 2013 (hereinafter API vs. EPA). The American Petroleum Institute (API) petitioned the court about two aspects of the RFS. The first aspect, which is directly related to cellulosic biofuels, was an objection to EPA's 2012 cellulosic biofuels projection. The court agreed with API that "because EPA's methodology for making its cellulosic biofuel projection did not take neutral aim at accuracy, it was an unreasonable exercise of agency discretion." The court went on to explain that EPA is tasked with establishing "a projection that aims at accuracy, not at deliberately indulging a greater risk of overshooting than undershooting." The 2012 rule was vacated and remanded to EPA, which set the final 2012 level at zero. The second aspect of the petition was an objection to EPA not providing a reasoned explanation for its refusal to reduce the 2012 advanced biofuels volume standard. The court rejected this part of the petition.

6 U.S. Environmental Protection Agency, "Regulation of Fuels and Fuel Additives: 2013 Renewable Fuel Standards; Final Rule," 78 Federal Register, August 15, 2013.

7 EPA does not state the actual volume amount in the proposed rule. CRS analysis suggests that the actual volume will be less than 10 million gallons. U.S. Environmental Protection Agency, "2014 Standards for the Renewable Fuel Standard Program; Proposed Rule," 78 Federal Register, November 29, 2013.

8 For more information on cellulosic biofuels, see CRS Report RL34738, Cellulosic Biofuels: Analysis of Policy Issues for Congress.

9 For more information on ethanol, see CRS Report RL33290, Fuel Ethanol: Background and Public Policy Issues.

10 Cellulose feedstocks can also be used to provide heat or generate electricity via gasification, combustion, anaerobic digestion, and other conversion processes. For more information, see CRS Report R40667, Anaerobic Digestion: Greenhouse Gas Emission Reduction and Energy Generation; and CRS Report R41440, Biomass Feedstocks for Biopower: Background and Selected Issues.

11 For more information on the RFS, see CRS Report R40155, Renewable Fuel Standard (RFS): Overview and Issues.

12 EPA reports that the first cellulosic biofuel RINs were registered in 2012 for cellulosic biofuel produced at demonstration-scale facilities. For 2009 and 2010, cellulosic production was registered under the RFS pre-EISA. Under the earlier definition, cellulosic biofuel could either be fuel produced from cellulosic feedstocks or fuel from conventional feedstocks using biomass for process energy. EISA eliminated the latter option.

13 U.S. Dept. of Energy, U.S. Dept. of Agriculture, Biomass as Feedstock for a Bioenergy and Bioproducts Industry: The Technical Feasibility of a Billion-Ton Annual Supply, April 2005, http://www1.eere.energy.gov/biomass/pdfs/ final_billionton_vision_report2.pdf.

14 Diane Greer, "Creating Cellulosic Ethanol: Spinning Straw into Fuel," BioCycle, April 2005; Biotechnology Industry Organization, Achieving Sustainable Production of Agricultural Biomass for Biorefinery Feedstock, Washington, DC, 2006, http://www.bio.org/ind/biofuel/ SustainableBiomassReport.pdf; and Biotechnology Industry Organization, "Energy Bill

Biofuels Mandates Will Be Achievable with Biotechnology Advances," press release, November 18, 2007, http://bio.org/news/pressreleases/newsitem.asp?id= 2007_1218_01& p=yes.

[15] Ian Talley, "Renewed Energy: US Biofuel Mandate Calls for Big Production Boost," Dow Jones International News, December 18, 2007, at http://www.factiva.com/.

[16] For more information on EPA's waiver authority, see CRS Report RS22870, Waiver Authority Under the Renewable Fuel Standard (RFS).

[17] 42 U.S.C. 7545(o)(7)(D).

[18] In its 2014 Standards for the Renewable Fuel Standard Program proposed rule, EPA reported it tracked the progress of several dozen potential cellulosic biofuel production facilities. U.S. Environmental Protection Agency, "2014 Standards for the Renewable Fuel Standard Program; Proposed Rule," 78 Federal Register, November 29, 2013.

[19] The original RFS established by §1501 of EPAct05 required 4.0 billion gallons of renewable fuel for 2006, ascending to 7.5 billion gallons by 2012. The original RFS would have required that 250 million gallons of the renewable fuel be derived from cellulosic biomass starting in 2013; whereas the expanded RFS, or RFS2, required 250 million gallons of cellulosic biofuel in 2011.

[20] E-mail from Bryan Stockton of ML Strategies, December 13, 2012.

[21] E-mail from Kim Nelson of American Process, Inc., December 11, 2012.

[22] Clean Fuels Development Corporation, Nebraska Ethanol Board, and U.S. Dept. of Agriculture, A Guide for Evaluating the Requirements of Ethanol Plants, 2006, http:// www.ethanol.org/pdf/contentmgmt/ guide_for_evaluating_the_requirements_of_ethanol_plants.pdf.

[23] For more information on federal spending for cellulosic biofuels, see CRS Report RL34738, Cellulosic Biofuels: Analysis of Policy Issues for Congress.

[24] For more information on biomass definitions, see CRS Report R40529, Biomass: Comparison of Definitions in Legislation.

[25] University of Tennessee, Growing and Harvesting Switchgrass for Ethanol Production in Tennessee, SP701-A, http://www.utextension.utk.edu/publications/spfiles/SP701-A.pdf.

[26] U.S. Environmental Protection Agency, "Regulation of Fuels and Fuel Additives: 2013 Renewable Fuel Standards; Final Rule," 78 Federal Register, August 15, 2013.

[27] A biorefinery is a facility that processes biomass into biofuels.

[28] One company, POET, has devised a feedstock supply plan whereby the feedstock would be stored by the producer, under a four-year contract, and delivered to the biorefinery as needed. Jim Lane, "Should the Renewable Fuel Standard be Scrapped, or Revised?" Biofuels Digest, November 2, 2011, http://biofuelsdigest.com/bdigest/2011/11/01/are-you-any-less-worried-about-energy-security-today-than-you-were-4-years-ago-asks-bp/.

[29] Several studies have estimated the current or long-term cellulosic feedstock supply available for conversion to biofuels on a national basis. The estimates provided in each study are based on numerous assumptions and modeling techniques unique to each study. DOE studies include the U.S. Billion-Ton Update: Biomass Supply for a Bioenergy and Bioproducts Industry, ORNL/TM-2011/224, August 2011; and Biomass as Feedstock for a Bioenergy and Bioproducts Industry: The Technical Feasibility of a Billion-Ton Annual Supply, April 2005. Non-DOE studies include the National Academy of Sciences, Renewable Fuel Standard: Potential Economic and Environmental Effects of U.S. Biofuel Policy, 2011; National Academy of Sciences, National Academy of Engineering, National Research Council, Liquid Transportation Fuels from Coal and Biomass: Technological Status, Costs, and Environmental Impacts, Washington, DC, 2009; and Biomass Research

and Development Initiative, Increasing Feedstock Production for Biofuels Economic Drivers, Environmental Implications, and the Role of Research, December 2008.

[30] For example, USDA has $181 million available to fund, through the Biorefinery Assistance Program, commercial-scale biorefineries or to retrofit existing facilities with appropriate technology to develop advanced biofuels. U.S. Department of Agriculture, "USDA Announces Availability of Funding to Develop Advanced Biofuels Projects," press release, October 21, 2013, http://www.usda.gov/wps/portal/usda/usdahome? contentid=2013/10/0195.xml.

[31] For example, the DOE awarded $10 million for five advanced biofuel and bio-based product projects. U.S. Department of Energy, Energy Department Awards $10 Million to Develop Advanced Biofuels and Bio-based Products, January 3, 2013, http://apps1.eere.energy.gov/news/progress_alerts.cfm/pa_id=822.

[32] A loan guarantee is defined as a "pledge with respect to the payment of all or a part of the principal or interest on any debt obligation of a non-federal borrower to a non-federal lender." The LGP was first authorized under Title XVII of EPAct05 and then amended under the American Recovery and Reinvestment Act of 2009 (P.L. 111-5). DOE may issue loan guarantees to eligible projects that "avoid, reduce, or sequester air pollutants or anthropogenic emissions of greenhouse gases" and "employ new or significantly improved technologies as compared to technologies in service in the United States at the time the guarantee is issued." Eligible projects include commercial-scale renewable energy systems. EISA authorized the DOE to issue loan guarantees in part to support renewable energy projects. For more information on loan guarantees, see CRS Report R42152, Loan Guarantees for Clean Energy Technologies: Goals, Concerns, and Policy Options.

[33] In 2011, a Section 1705 loan for $132.4 million was awarded to Abengoa Bioenergy Biomass of Kansas, LLC for a commercial-scale cellulosic ethanol project. U.S. Department of Energy, "Energy Department Finalizes $132 Million Loan Guarantee to Support the Abengoa Bioenergy Project," press release, September 29, 2011, http://energy.gov/articles/energy-department-finalizes-132-million-loan-guarantee-support-abengoa-bioenergy-project.

[34] Renewable Fuels Association, "2010 State of the Industry Address," 2010 National Ethanol Conference, Orlando, FL, February 16, 2010, http://ethanolrfa.3cdn.net/b76292e4bf133edd34_e1m6bhh33.pdf.

[35] U.S. Government Accountability Office, Further Actions Are Needed to Improve DOE's Ability to Evaluate and Implement the Loan Guarantee Program, GAO-10-627, July 2010, http://www.gao.gov/new.items/d10627.pdf; U.S. Government Accountability Office, DOE Loan Guarantees: Further Actions Are Needed to Improve Tracking and Review of Applications, GAO-12-157, March 2012, http://www.gao.gov/assets/590/589210.pdf.

[36] For more information, see CRS Report R41985, Renewable Energy Programs and the Farm Bill: Status and Issues.

[37] Email from Ashley Martin, USDA, March 1, 2013.

[38] BCAP receives its authorization from Title IX of the Farm Security and Rural Investment Act of 2002 (P.L. 107-171) and was amended by Title IX of the Food, Conservation, and Energy Act of 2008 (P.L. 110-246). For more information on BCAP, see CRS Report R41296, Biomass Crop Assistance Program (BCAP): Status and Issues; or visit http://www.fsa.usda.gov/FSA/webapp?area=home&subject=ener&topic=bcap.

[39] This program's funding, and other energy programs' funding in the farm bill, depends on the reauthorization of the 2013 farm bill. For more information, see CRS Report R43076, The

2013 Farm Bill: A Comparison of the Senate-Passed (S. 954) and House-Passed (H.R. 2642, H.R. 3102) Bills with Current Law.

[40] BCAP project areas are specific geographic areas where producers may enroll land to grow specified biomass crops. Participants may be eligible to receive financial and technical assistance as well as annual payments to establish these crops.

[41] Matching payments are provided for the collection, harvest, storage and transport of eligible biomass materials to qualified biomass conversion facilities.

[42] Email from Ashley Martin, USDA, March 1, 2013.

[43] For more information, see CRS Report R42566, Alternative Fuel and Advanced Vehicle Technology Incentives: A Summary of Federal Programs.

[44] The energy content of renewable fuels differ. There are times when an ethanol-equivalence value for a renewable fuel with a higher energy content than ethanol may allow for that fuel to count more per gallon towards a mandate than other biofuels.

[45] U.S. Environmental Protection Agency, "Regulation of Fuels and Fuel Additives: Changes to Renewable Fuel Standard Program; Final Rule," 75 Federal Register, March 26, 2010.

[46] U.S. Environmental Protection Agency, "Regulation of Fuels and Fuel Additives: 2011 Renewable Fuel Standards; Final Rule," Federal Register, December 9, 2010.

[47] Environmental Protection Agency, Regulation of Fuels and Fuel Additives: 2012 Renewable Fuel Standards, Final Rule, December 22, 2011. This rule was vacated by the U.S. Court of Appeals for the D.C. Circuit in API v. EPA. EPA revised the 2012 rule and set the final required level at zero.

[48] U.S. Environmental Protection Agency, "Regulation of Fuels and Fuel Additives: 2013 Renewable Fuel Standards; Final Rule," 78 Federal Register, August 6, 2013.

[49] U.S. Environmental Protection Agency, "2014 Standards for the Renewable Fuel Standard Program; Proposed Rule," 78 Federal Register, November 29, 2013.

[50] For more information on EMTS, see http://www.epa.gov/otaq/fuels/rfsdata/2010emts.htm.

[51] Concerns have been raised about fraudulent RINs, especially due to recent biodiesel fraudulent RIN activity. While RIN fraud is possible as discussed in CRS Report R42824, Analysis of Renewable Identification Numbers (RINs) in the Renewable Fuel Standard (RFS), there has been no reported fraudulent cellulosic biofuel RIN activity. EPA issued a proposed rule to establish a voluntary quality assurance program for verifying the validity of RINs. For more information, see U.S. Environmental Protection agency, RFS Renewable Identification Number (RIN) Quality Assurance Program, January 31, 2013, http://www.epa.gov/otaq/fuels/renewablefuels/documents/qap-nprm-pepub.pdf.

[52] 42 USC 7545 (o)(7)(D)(ii)

[53] U.S. Environmental Protection Agency, Questions and Answers on Changes to the Renewable Fuel Standard Program (RFS2), February 2013.

[54] Waiver credits are not allowed to be traded or banked for future use, and are only allowed to be used to meet the cellulosic biofuel standard for the year that they are offered.

[55] The original RFS (RFS1) did not require a cellulosic biofuel production volume until 2013, when 250 million gallons were mandated. Further, the definition of cellulosic biofuel was amended by EISA.

[56] EISA did not stipulate a 2.5-1 ratio for cellulosic biofuel for the RFS2. Instead, the ethanol equivalence value depends on the type of biofuel being produced.

[57] EMTS reports zero RINs available for 2011 cellulosic biofuel and cellulosic diesel.

[58] U.S. Environmental Protection Agency, "Regulation of Fuels and Fuel Additives: 2013 Renewable Fuel Standards; Proposed Rule," 78 Federal Register, February 7, 2013. The design of the RFS2 transition program allowed for RFS1 cellulosic RINs to be used towards

compliance under the RFS2 program. The RFS1 cellulosic RINs likely came from non-cellulosic feedstocks that qualified under the then-effective definition of cellulosic biomass. EPA reports the 2010 cellulosic biofuel standard could have been met with cellulosic biofuel RINs generated under RFS2 regulations after July 1, 2010, cellulosic biomass ethanol RINs generated under RFS1 regulations between January 1, 2010 and July 1, 2010, and cellulosic biomass ethanol RINs generated under RFS1 regulations in 2009. U.S. Environmental Protection Agency, Questions and Answers on Changes to the Renewable Fuel Standard Program (RFS2), http://www.epa.gov/otaq/fuels/renewablefuels/ compliance help/rfs2-aq.htm.

[59] See http://www. epa.gov/otaq/fuels/rfsdata/rfs2cellulosicwaivercredits.htm.

[60] EMTS reports zero RINs available for 2011 cellulosic biofuel and cellulosic diesel. For more information, see http://www.epa.gov/otaq/fuels/rfsdata/2011emts.htm.

[61] The statute allows companies to purchase cellulosic biofuel waiver credits in lieu of submitting RINs in years when EPA lowers the mandate. U.S. Environmental Protection Agency, "Regulation of Fuels and Fuel Additives: 2013 Renewable Fuel Standards; Final Rule," 78 Federal Register, August 6, 2013.

[62] See http://www.epa.gov/otaq/fuels/rfsdata/rfs2cellulosicwaivercredits.htm. If the 2011 cellulosic biofuel standard is rescinded, EPA reports that it would refund the money paid by obligated parties to purchase cellulosic waiver credits to comply with the 2011 cellulosic biofuel standard.

[63] U.S. Environmental Protection Agency, Enviroflash, Update - 2012 Cellulosic Biofuel Standard Mandate Issued, February 27, 2013.

[64] Blue Sugar Corporation is the first company to be issued cellulosic biofuel RINs. EPA reports these RINs were "retired" because the cellulosic ethanol produced was exported to Brazil. For more information, see http://www.epa.gov/otaq/fuels/rfsdata/2012emts.htm. U.S. Environmental Protection Agency, "Regulation of Fuels and Fuel Additives: 2013 Renewable Fuel Standards; Final Rule," 78 Federal Register, August 15, 2013.

[65] EIA first estimated that 9.6 million actual gallons of cellulosic biofuel would be available in 2013. Approximately six months later, EIA provided EPA with a revised projection of 4 million actual gallons of cellulosic biofuel. U.S. Environmental Protection Agency, "Regulation of Fuels and Fuel Additives: 2013 Renewable Fuel Standards; Final Rule," 78 Federal Register, August 15, 2013.

[66] U.S. Environmental Protection Agency, "Regulation of Fuels and Fuel Additives: 2013 Renewable Fuel Standards; Final Rule," 78 Federal Register, August 15, 2013.

[67] INEOS Bio announced on July 31, 2013, that it started commercial production of cellulosic ethanol. KiOR announced on August 8, 2013, that it shipped over 75,000 gallons of cellulosic fuel from its commercial facility.

[68] John M. Biers, "Non-Food Ethanol Projects Coming to Life," Dow Jones Newswires, November 20, 2012.

[69] "Zero Dark Ethanol," The Wall Street Journal, January 30, 2013.

[70] "Oil Industry Sues EPA A Second Time Over RFS Mandate," The Energy Daily, July 26, 2012.

[71] U.S. Environmental Protection Agency, "Regulation of Fuels and Fuel Additives: 2013 Renewable Fuel Standards; Final Rule," 78 Federal Register, August 15, 2013. Many of the concerns raised about RIN acquisition are unrelated to cellulosic biofuels and are outside the scope of this report. For more information on compliance with the 2013 RFS standards generally, see CRS Report R40155, Renewable Fuel Standard (RFS): Overview and Issues;

and CRS Report R42824, Analysis of Renewable Identification Numbers (RINs) in the Renewable Fuel Standard (RFS).

[72] 42 U.S.C. 7545 (o)(7)(F).

[73] Some may argue that there is evidence of this currently occurring, as 2014 is the first year that EPA proposed to lower both the advanced biofuel mandate and the total renewable fuel mandate, partly due to an "inadequate domestic supply of these fuels." U.S. Environmental Protection Agency, "2014 Standards for the Renewable Fuel Standard Program; Proposed Rule," 78 Federal Register, November 29, 2013.

[74] U.S. Environmental Protection Agency, "Regulation of Fuels and Fuel Additives: 2013 Renewable Fuel Standards; Final Rule," 78 Federal Register, August 15, 2013.

[75] U.S. Environmental Protection Agency, "2014 Standards for the Renewable Fuel Standard Program; Proposed Rule," 78 Federal Register, November 29, 2013.

[76] Wallace E. Tyner and Sarah Brechbill, "Cellulosic Biofuels: Feedstocks, Conversion Technologies, Economics, and Policy Issues," CRS Workshop on the Development of the U.S. Cellulosic Biofuels Industry, Washington, DC, October 6, 2009; and conversation with Wallace Tyner from Purdue University, February 2, 2010; CRS Report R41460, Cellulosic Ethanol: Feedstocks, Conversion Technologies, Economics, and Policy Options, by Randy Schnepf; EPA, Renewable Fuel Standard Program (RFS2) Regulatory Impact Analysis, EPA-420-R-10-006, Washington, DC, February 2010, pp. 171 and 186, http://www.epa.gov/oms/renewablefuels/420r10006.pdf.

[77] Advanced Ethanol Council (AEC), Cellulosic Biofuels Industry Progress Report 2012-2013, December 2012, http://ethanolrfa.org/page/-/PDFs/AEC%20Cellulosic%20Biofuels% 20 Industry%20Progress%20Report%202012- 2013.pdf?nocdn=1.' AEC acknowledges the report does not profile all cellulosic biofuel projects under development in the United States.

[78] Environmental Entrepreneurs, Advanced Biofuel Market Report 2013: Capacity Through 2016, 2013.

[79] EIA reports that "its forecasts and projections to date have proven to be too optimistic, as volumes have been below expectations." U.S. Energy Information Administration, Cellulosic biofuels begin to flow but in lower volumes than foreseen by statutory targets, February 26, 2013.

[80] The blend wall is the upper limit to the total amount of ethanol that can be blended into U.S. gasoline without exceeding the currently utilized 10% level. For more information, see CRS Report R40445, Intermediate-Level Blends of Ethanol in Gasoline, and the Ethanol "Blend Wall."

[81] Keith Johnson, "Navy's Biofuel Plan Gets Senate Support," The Wall Street Journal, November 28, 2012; P.L. 112- 81; H.Rept. 112-329.

[82] U.S. Environmental Protection Agency, "2014 Standards for the Renewable Fuel Standard Program; Proposed Rule," 78 Federal Register, November 29, 2013.

[83] Environmental Entrepreneurs, Advanced Biofuel Market Report 2013: Capacity Through 2016. CRS tallied federal funding from Appendix H of the report for those companies with facilities coming online in the United States before 2016 that were identified as producing the fuel type cellulosic ethanol in Appendix A of the report.

[84] POET LLC, "POET and Agrivida sign technology collaboration joint development agreement," press release, August 21, 2012.

[85] U.S. Environmental Protection Agency, Regulation of Fuels and Fuel Additives: Identification of Additional Qualifying Renewable Fuel Pathways under the Renewable Fuel Standard Program, February 22, 2013.

[86] U.S. Energy Information Administration, Cellulosic biofuels begin to flow but in lower volumes than foreseen by statutory targets, February 26, 2013.

[87] Although some stakeholders have questioned whether EPA does, in fact, use the best available evidence.

[88] U.S. Environmental Protection Agency, Enviroflash, Update - 2012 Cellulosic Biofuel Standard Mandate Issued, February 27, 2013; U.S. Environmental Protection Agency, "2014 Standards for the Renewable Fuel Standard Program; Proposed Rule," 78 Federal Register, November 29, 2013.

[89] American Petroleum Institute v. EPA, 706 F.3d 474 (D.C. Cir. 2013).

[90] EPA reports that the INEOS Bio, KiOR, Poet, and Abengoa facilities all have feedstock agreements or contracts in place.

In: The Renewable Fuel Standard
Editor: Lydia Rider
ISBN: 978-1-63321-829-1

Chapter 5

THE RENEWABLE FUEL STANDARD: ISSUES FOR 2014 AND BEYOND*

Terry Dinan, Ron Gecan and David Austin

Note: Numbers in the text and tables of this report may not add up to totals because of rounding.

SUMMARY

The Renewable Fuel Standard (RFS) establishes minimum volumes of various types of renewable fuels that must be included in the United States' supply of fuel for transportation. Those volumes—as defined by the Energy Independence and Security Act of 2007 (EISA)— are intended to grow each year through 2022. In recent years, the requirements of the RFS have been met largely by blending gasoline with ethanol made from cornstarch. In the future, EISA requires the use of increasingly large amounts of "advanced biofuels," which include diesel made from biomass (such as soybean oil or animal fat), ethanol made from sugarcane, and cellulosic biofuels (made from converting the cellulose in plant materials into fuel).

One of the main goals of the Renewable Fuel Standard is to reduce U.S. emissions of greenhouse gases, which contribute to climate change.

* This is an edited, reformatted and augmented version of a report released by the Congressional Budget Office, June 2014.

EISA requires that the emissions associated with a gallon of renewable fuel be at least a certain percentage lower than the emissions associated with the gasoline or diesel that the renewable fuel replaces. Advanced biofuels and the subcategory of cellulosic biofuels are required to meet more stringent emission standards than those that apply to corn ethanol.

Policymakers and analysts have raised concerns about the RFS, particularly about the feasibility of complying with the standard, whether it will increase prices for food and transportation fuels, and whether it will lead to the intended reductions in greenhouse gas emissions. Because of those concerns, some policymakers have proposed repealing or revising the Renewable Fuel Standard.

In this analysis, the Congressional Budget Office (CBO) evaluates how much the supply of various types of renewable fuels would have to increase over the next several years to comply with the RFS. CBO also examines how food prices, fuel prices, and emissions would vary in an illustrative year, 2017, under three scenarios for the Renewable Fuel Standard:

- *The EISA volumes scenario*, in which fuel suppliers would have to meet the total requirement for renewable fuels, the requirement for advanced biofuels, and the cap on corn ethanol that are stated in EISA for 2017—but not the requirement for cellulosic biofuels, because the capacity to produce enough of those fuels is unlikely to exist by 2017;
- *The 2014 volumes scenario,* in which the Environmental Protection Agency (EPA)—which has some discretion to modify the mandates of EISA—would keep the RFS requirements for the next several years at the same amounts it has proposed for 2014; and
- *The repeal scenario*, in which lawmakers would immediately abolish the RFS.

The repeal scenario would require Congressional action. In the absence of such action (or of legal restrictions), CBO considers the 2014 volumes scenario much more likely than the EISA volumes scenario, which would require a large and rapid increase in the use of advanced biofuels and would cause the total percentage of ethanol in the nation's gasoline supply to rise to levels that would require significant changes in the infrastructure of fueling stations.

Full Compliance with the Mandates in EISA Poses Significant Challenges

The rising requirements in EISA would be very hard to meet in future years because of two main obstacles, which relate to the supply of cellulosic biofuels and the amount of ethanol that older vehicles are said to be able to tolerate. Fuel suppliers have had trouble meeting the annual requirements for cellulosic biofuels because making such fuels is complex, capital-intensive, and costly. Although production capacity is expanding, only a few production facilities are currently operating. The industry's capacity in coming years is projected to fall far short of what would be necessary to achieve the very rapid growth in the use of cellulosic biofuels required by EISA.

Ethanol is the most common form of renewable fuel; however, adding increasing volumes of it to the U.S. fuel supply could be difficult. Currently, most gasoline sold in the United States is actually a blend (referred to as E10) that contains up to 10 percent ethanol—the maximum concentration that is feasible to avoid corrosion damage to the fuel systems of older vehicles. EISA's growing requirements for the total gallons of renewable fuels to be used each year, combined with a projected decline in gasoline use, suggest that the average concentration of ethanol in gasoline would have to rise to well above that 10 percent "blend wall," potentially increasing to about 25 percent by 2022. More ethanol could be accommodated in the fuel supply if motorists who drive "flex-fuel" vehicles, which can run on blends that contain as much as 85 percent ethanol (referred to as E85), bought larger amounts of such fuel. But at present, fewer than 2 percent of filling stations in the United States sell high-ethanol blends. Given the design of the RFS, the cost of encouraging additional sales of high-ethanol fuel falls on the producers and consumers of gasoline and diesel.

Because of the challenges described above, EPA has eliminated or greatly reduced the annual requirements for cellulosic biofuels in the RFS in past years. For 2014, EPA has also proposed regulations that would reduce the requirements for advanced biofuels and for total renewable fuels, in recognition of the difficulties posed by the blend wall. Although scaling back those standards addresses existing compliance problems and decreases compliance costs in the short run, it also reduces incentives for companies to invest in production capacity for cellulosic and other advanced biofuels and to expand the availability of high-ethanol blends.

Meeting the Total Volumes of Advanced Biofuels Specified in EISA Would Require Extremely Large Increases in the Production of Those Fuels

For the scenario in which fuel suppliers would have to comply with the total volumes of advanced biofuels and of renewable fuels as a whole stated in EISA, CBO assumed that EPA would allow suppliers to substitute other forms of advanced biofuels for cellulosic biofuels, as it has done in the past. Fuel suppliers would most likely do so using two types of advanced biofuels: biomass-based diesel (mostly produced in the United States) and sugarcane ethanol (nearly all imported from Brazil). However, relying on that strategy for 2017 would necessitate extremely large increases in the production of those fuels: for example, more than a 100 percent rise in U.S. production of biomass-based diesel and more than a 45 percent increase in Brazil's production of sugarcane ethanol.

Food Prices Would Be Similar Whether the RFS Was Continued or Repealed

Roughly 40 percent of the U.S. corn supply is used to make ethanol. To the extent that the Renewable Fuel Standard increases the demand for corn ethanol, it will raise corn prices and put upward pressure on the prices of foods that are made with corn—ranging from corn-syrup sweeteners to meat, poultry, and dairy products. CBO expects that roughly the same amount of corn ethanol would be used in 2017 if fuel suppliers had to meet requirements equal to EPA's proposed 2014 volumes or if lawmakers repealed the RFS, because suppliers would probably find it cost-effective to use a roughly 10 percent blend of corn ethanol in gasoline in 2017 even in the absence of the RFS. Therefore, food prices would also be about the same under the 2014 volumes scenario and the repeal scenario.

By contrast, corn ethanol use in 2017 would be about 15 percent (or 2 billion gallons) higher under the EISA volumes scenario. CBO estimates that the resulting increase in the demand for corn would raise the average price of corn by about 6 percent. However, because corn and food made with corn account for only a small fraction of total U.S. spending on food, that total spending would increase by about one-quarter of one percent.

Meeting the Total Volumes of Advanced Biofuels Specified in EISA Would Have Significant Effects on Prices of Transportation Fuels

Because fuel suppliers would be likely to use roughly a 10 percent blend of corn ethanol in gasoline in 2017 even without the RFS, the overall use of renewable fuels in that year would be very similar under the 2014 volumes scenario and under the repeal scenario, CBO estimates. Consequently, prices of transportation fuels would probably be roughly the same in those two cases.

Under the EISA volumes scenario, however, fuel suppliers would have to use more than three times as many gallons of advanced biofuels, and they would have to add much more ethanol to the gasoline supply than could be accommodated by selling only a 10 percent blend. (Under all of the scenarios, CBO anticipates that EPA would sharply reduce the requirement for cellulosic biofuels, given the limited production capacity for those fuels expected to exist in 2017.) Using a range of estimates of the price premium necessary to encourage sufficient additional supplies of advanced biofuels and the price subsidy necessary to motivate sufficient sales of E85, CBO estimates that complying with the EISA volumes scenario would have the following effects on the prices of three key types of transportation fuels in 2017:

- The price of petroleum-based diesel would rise by 30 cents to 51 cents per gallon, or 9 percent to 14 percent (because the RFS requires fuel suppliers to bear the cost of ensuring that certain amounts of renewable fuels are used for each gallon of petroleum-based fuel that they sell);
- The price of E10—which is currently the most commonly used transportation fuel in the United States— would increase by 13 cents to 26 cents per gallon, or 4 percent to 9 percent; and,
- The price of E85 would decline by 91 cents to $1.27 per gallon, or 37 percent to 51 percent.

Because the changes in the production and use of renewable fuels required under the EISA volumes scenario are so large—and because little information is available about how the supply of and demand for renewable fuels respond to changes in their price—those estimates are highly uncertain. Actual price changes could fall outside the ranges described above.

Reductions in Greenhouse Gas Emissions Because of the RFS Would Be Small in the Near Term but Could Be Larger Over the Long Term

The production and use of different types of renewable fuels involve different amounts of greenhouse gas emissions. Estimates of those emissions are uncertain, and researchers' predictions vary considerably. However, available evidence suggests that replacing gasoline with corn ethanol has only limited potential for reducing emissions (and some studies indicate that it could increase emissions). The success of the RFS in reducing the emissions from transportation fuels will depend mainly on the extent to which it causes people to substitute advanced biofuels—particularly cellulosic biofuels—for gasoline or diesel over the long run. However, a trade-off exists between the goal of limiting the cost of complying with the RFS (for example, by reducing the requirements for cellulosic biofuels) and the goal of providing a strong incentive for the development of better technologies for advanced biofuels.

Overview of the Renewable Fuel Standard and Its Implementation

Lawmakers enacted the Renewable Fuel Standard in 2005 and expanded its requirements in 2007 in the Energy Independence and Security Act. The standard is imposed on suppliers (generally refiners or importers) of gasoline and diesel fuels used for transportation. It aims to foster greater use of fuels made from plants, plant products, and other renewable sources, thereby reducing the United States' dependence on petroleum and the greenhouse gas emissions from fuel use. The Environmental Protection Agency is charged with implementing the standard and ensuring compliance.

What the RFS Requires

The Energy Independence and Security Act sets minimum volumes of renewable fuels that suppliers must blend into the nation's supply of transportation fuel each year.

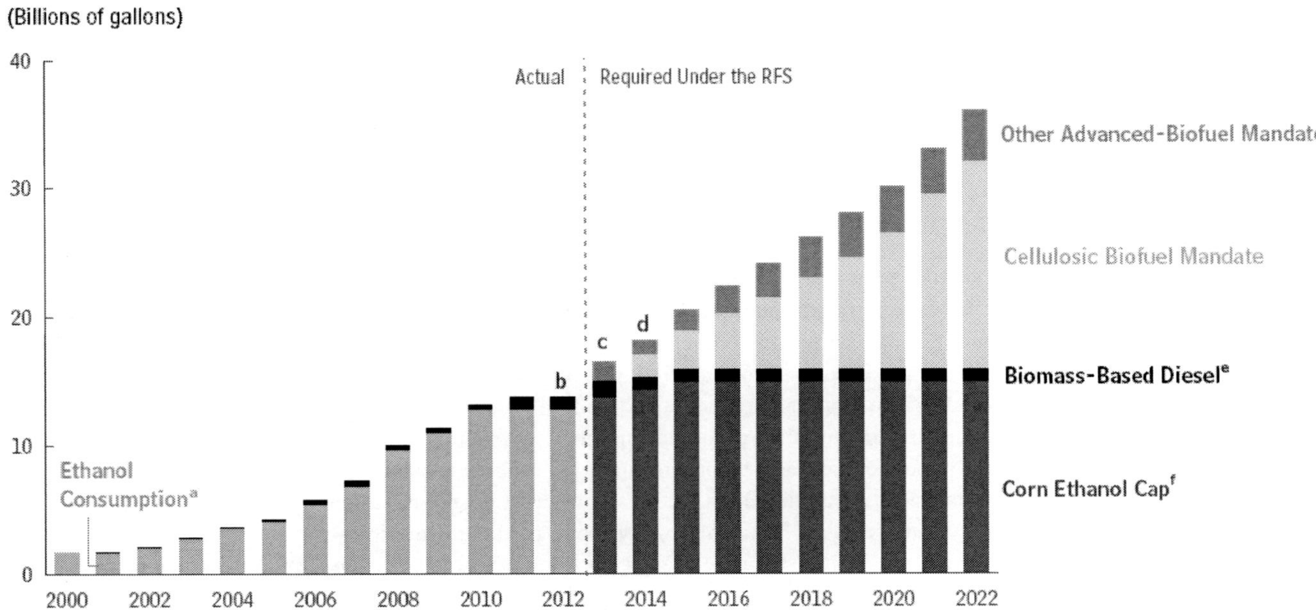

Source: Congressional Budget Office based on data for 2000 to 2012 from Energy Information Administration, *Monthly Energy Review,* DOE/EIA-0035(2014/04) (April 2014), www.eia.gov/totalenergy/data/monthly, and requirements for 2013 to 2022 from the Energy Independence and Security Act of 2007.

Note: RFS = Renewable Fuel Standard.

[a] Most of the ethanol used in the United States in the past consisted of corn ethanol, although relatively small amounts of sugarcane ethanol and other types of advanced biofuels, either produced domestically or imported, were also used.

[b] Because of high corn prices in 2012, use of renewable fuels was about the same in that year as in 2011. That use was less than the amounts mandated for 2012, but fuel blenders and importers achieved compliance with the RFS by submitting "renewable identification numbers" (or RINs) that they had accumulated from exceeding their obligations in prior years.

[c] For 2013, the Energy Independence and Security Act (EISA) originally required the use of 1 billion gallons each of cellulosic biofuels and biomass-based diesel. In August 2013, the Environmental Protection Agency (EPA) retroactively reduced the cellulosic biofuel requirement for that year to 6 million gallons and raised the mandate for biomass-based diesel to 1.28 billion gallons. Complete data on the actual use of renewable fuels in 2013 were not yet available when this report was published.

[d] The amounts shown here for 2014 are those required under EISA. However, EPA has proposed reducing the 2014 requirement for cellulosic biofuels from 1.75 billion gallons to 17 million gallons, the requirement for advanced biofuels from 3.75 billion gallons to 2.2 billion gallons, and the cap on the amount of corn ethanol that can be used to meet the total requirement for renewable fuels from 14.4 billion gallons to 13.0 billion gallons. EPA has also proposed increasing the requirement for biomass-based diesel from 1 billion gallons to 1.28 billion gallons. Under those proposals, the total requirement for renewable fuels in 2014 would decline from 18.15 billion gallons to 15.21 billion gallons, compared with 16.55 billion gallons in 2013.

[e] The amounts of biomass-based diesel shown here for 2014 and later years reflect the minimum requirement of 1 billion gallons specified in EISA. EPA will set the actual requirement for each year through future rulemaking.

[f] The cap on corn ethanol represents the maximum amount of such ethanol that can used to meet the total requirement for renewable fuels under EISA.

Figure 1. Past Use of Renewable Fuels and Future Requirements of the Renewable Fuel Standard.

Except for corn ethanol made in certain facilities, the renewable fuels used to comply with the RFS must be certified by EPA as having greenhouse gas emissions that are at least 20 percent lower than the emissions associated with the fuels that they replace. Corn ethanol made at plants built or under construction before December 20, 2007, is exempt from the emission requirements, which means that much of the corn ethanol used today does not need to have lower emissions than gasoline. The total minimum volume of renewable fuels specified in EISA rises each year through 2022 (see Figure 1) and includes the requirement that an increasing share of that volume be met with advanced biofuels, which must have greenhouse gas emissions that are at least 50 percent lower than those of conventional fuels.

So far, fuel suppliers have been able to comply with the RFS largely by blending gasoline with corn ethanol, which is made from the starch in corn kernels. By 2022, EISA requires the use of 36 billion gallons of renewable fuels. Of those, at least 21 billion gallons must be advanced biofuels, including the following:

- At least 16 billion gallons of cellulosic biofuels, which are made from the cellulose in various plant materials, including grasses and corn stover (the residue left after corn is harvested). Cellulosic biofuels must have emissions that are at least 60 percent lower than their petroleum-based counterparts.
- At least 1 billion gallons of biomass-based diesel (typically made from soybean or other vegetable oils). EPA has the discretion to set the mandate for biomass-based diesel at a higher level.[1]

The other 4 billion gallons (or less) can consist of any type of advanced biofuel that meets the 50-percent-lower emission standard, such as noncellulosic ethanol made from sugarcane.

The portion of the RFS that does not have to be met with advanced biofuels—in 2022, up to 15 billion gallons— can be met with other qualifying renewable fuels, such as corn ethanol. Thus, the requirements for cellulosic biofuels and for biomass-based diesel are nested within the requirement for advanced biofuels, which in turn is nested within the overall requirement for renewable fuels.[2]

The total volume of renewable fuels mandated by EISA increases much faster than the projected growth in the use of gasoline and diesel. As a result, under the RFS, renewable fuels would make up a greater share of the U.S.

supply of transportation fuel over time, rising from about 7 percent in 2013 to about 18 percent in 2022 (see Figure 2).

How EPA Implements the RFS

To ensure that fuel suppliers use the mandated volumes of renewable fuels, the Environmental Protection Agency translates the yearly volume requirements in EISA into percentage standards (sometimes called blend requirements) that are based on projections of the total amount of gasoline and diesel that will be used in that year. For example, if the projected amount was 100 billion gallons and the total renewable fuel requirement was 14 billion gallons, EPA would set a 14 percent blend requirement.

Further, if the nested mandates for advanced biofuels and for biomass-based diesel were 4 billion gallons and 2 billion gallons, respectively, EPA would establish a 4 percent blend requirement for advanced biofuels and a 2 percent requirement for biomass-based diesel.

To monitor suppliers' compliance with the requirements, EPA assigns a unique "renewable identification number" (RIN) to each qualifying gallon of renewable fuel. Every RIN includes a code that identifies which of the four RFS categories—total renewable fuels, advanced biofuels, cellulosic biofuels, or biomass-based diesel—the gallon satisfies. Each fuel supplier, regardless of what kind of fuel it produces or imports, must meet all of the blend requirements for a given compliance year. The supplier can do that by using the required amounts of renewable fuels itself and submitting the corresponding RINs to EPA to demonstrate compliance, by purchasing RINs from other suppliers that have excess RINs to sell, or by submitting RINs that it acquired in the previous year and saved for future use.[3] With the hypothetical requirements above, each fuel supplier would have to submit 14 RINs (including 4 for advanced biofuels and 2 for biomass-based diesel) for each 100 gallons of gasoline or diesel that it sold. Suppliers with excess biomass-based diesel RINs could either sell them or apply them toward their advanced-biofuel requirement.

EPA is also responsible for certifying that the types of renewable fuels used to comply with the RFS meet the emission requirements in EISA. Estimating the emissions associated with renewable fuels is difficult: Among other things, those emissions depend on the yields of the feedstock crops used to make the fuel, the amount of fertilizer used, the particular fuel-production technology employed, and changes in related factors (such as land use or fuel use inside or outside the United States) caused by the RFS. Predictions about those things are highly uncertain. Although EPA's predictions serve as the basis for determining whether a particular fuel can be used for complying with the RFS, actual emissions may differ from EPA's estimates. The final section of this report compares emission estimates from EPA and other researchers and discusses why they vary.

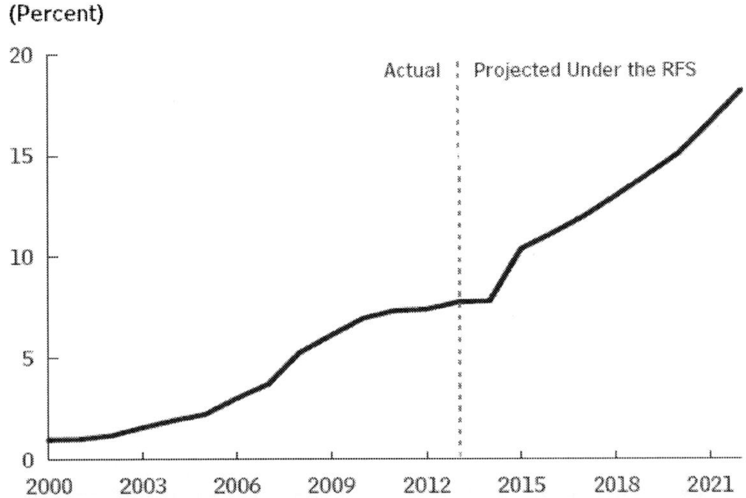

Source: Congressional Budget Office based on Energy Information Administration, *Annual Energy Outlook 2014, With Projections to 2040,* DOE/EIA-0383(2014) (April 2014), www.eia.gov/forecasts/aeo, and *Monthly Energy Review,* DOE/EIA-0035(2014/04) (April 2014), www.eia.gov/ totalenergy/data/monthly.

Notes: CBO's calculations are based on the Energy Information Administration's (EIA's) most recent projections of the use of blended gasoline and diesel fuel. EIA projects that in 2022, less renewable fuel will be used than required under the mandates specified in the Energy Independence and Security Act of 2007. CBO assumed that total use of renewable fuels would rise to the level mandated by that law, although total energy consumption of transportation fuels would remain the same. In addition, because part of the overall mandate for renewable fuels will be met with biomass-based diesel and the specific requirement for such diesel is set annually, CBO assumed that consumption of biomass-based diesel in future years would total either 1.28 billion gallons (the current requirement, which is equal to 1.92 billion compliance-equivalent gallons for the purposes of meeting the total mandate for renewable fuels), or 25 percent of the requirement for advanced biofuels, whichever is greater.

RFS = Renewable Fuel Standard.

Figure 2. Renewable Fuels as a Share of the Total U.S. Supply of Transportation Fuels.

Challenges in Meeting the Renewable Fuel Requirements of EISA

Complying with the Renewable Fuel Standard has raised several challenges, and EPA has modified the requirements of the RFS in past years in response to them. In particular, meeting the requirements for advanced biofuels specified in the Energy Independence and Security Act has posed two difficulties:

- The supply of cellulosic biofuels is limited because such fuels are complex and expensive to produce.
- The use of renewable fuels is constrained by a practical limit on the total amount of ethanol that can be blended into the fuel supply, given the technologies used by older vehicles and the existing fueling-station infrastructure. That limit was not a significant constraint in the past, but it is becoming one as the requirements of EISA increase and the use of transportation fuel grows more slowly than anticipated.

The way in which EPA has responded to those challenges has made it less costly for fuel suppliers to comply with the RFS. But at the same time, that response has lessened the incentives that the RFS provides for investment in renewable fuel infrastructure and for the development of improved technologies for producing advanced biofuels.

Limited Supply of Cellulosic Biofuels

To date, the greatest challenge in meeting the requirements specified in EISA has been the small supply of cellulosic biofuels. The industry that produces those fuels is in its infancy, and the volumes required by EISA far outstrip the projected growth in the industry's production capacity. EISA first set requirements for cellulosic biofuels in 2010, mandating the use of 100 million gallons in that year and larger amounts in each subsequent year. Before 2013, however, no commercial plants to produce cellulosic biofuels were in operation, and EPA virtually eliminated the requirements until that year.

Two commercial plants began making cellulosic biofuels in 2013, and more plants are expected to begin operating in 2014 and 2015. Even so, the gap between production capacity and the volumes of cellulosic biofuels mandated in EISA is expected to widen quickly. The Energy Information Administration forecasts that production of cellulosic biofuels will increase only to 327 million gallons by 2022, a small fraction of the 16 billion gallons required by EISA in that year (see Figure 3).[4]

Production capacity has been slow to expand for several reasons. Producing ethanol from cellulose is more complex than producing it from cornstarch, entails higher capital costs, and poses logistical problems. For example, commercial-scale use of cellulosic feedstocks requires that systems and equipment be developed to harvest the often-bulky materials and transport them to production facilities; for year-round production, seasonal feedstocks would also require ample storage space.

Difficulties in Using the Required Volume of Renewable Fuels

Ten percent is effectively the maximum ethanol content that blended fuel can contain and still be used by virtually all vehicles now on the road. That limit protects vehicles built before 2001, whose engines and fuel systems are thought to be vulnerable to corrosion from ethanol concentrations greater than 10 percent. For that reason, 10 percent constitutes a practical constraint, or blend wall, on how much ethanol most blended gasoline can accommodate. Many states limit ethanol concentrations to no more than 10 percent, except in fuels intended for flex-fuel vehicles, which can run on blends of as much as 85 percent ethanol.[5]

The challenges posed by the blend wall are expected to increase. When EISA was enacted, in 2007, use of blended gasoline in the United States totaled about 140 billion gallons a year and was projected to grow (see Figure 4). Thus, rising requirements for renewable fuels were not expected to raise concerns about the blend wall. Instead of growing, however, use of blended gasoline has declined slightly, to about 135 billion gallons a year, and the Energy Information Administration now projects that it will fall to about 125 billion gallons in 2022. (The agency's 2007 projection did not anticipate the decline in total annual vehicle-miles traveled and the increase in average fuel economy that have since occurred.)

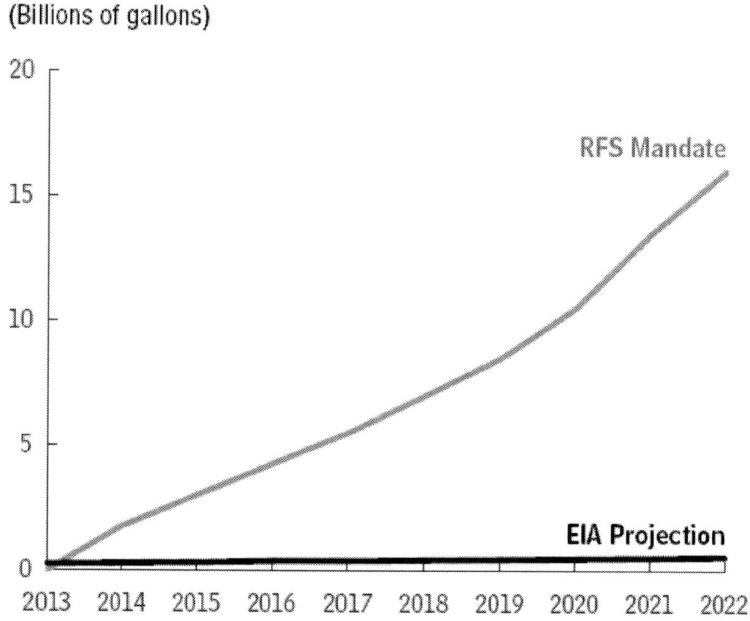

Source: Congressional Budget Office based on Energy Information Administration, *Annual Energy Outlook 2013, With Projections to 2040,* DOE/EIA-0383(2013) (April 2013), Figure 100, www.eia.gov/forecasts/archive/aeo13, and Regulation of Fuels and Fuel Additives: 2013 Renewable Fuel Standards, 78 Fed. Reg. 49794 (August 15, 2013), https://federalregister.gov/a/2013-19557.

Notes: The Energy Independence and Security Act of 2007 set annual requirements for cellulosic biofuels starting in 2010; however, the Environmental Protection Agency virtually eliminated the requirements before 2013 because of a lack of commercial production capacity for cellulosic biofuels.

RFS = Renewable Fuel Standard; EIA = Energy Information Administration.

Figure 3. Projected Use of Cellulosic Biofuels, Compared With the Use Mandated by the Renewable Fuel Standard.

If the latest projections prove accurate, the renewable fuel requirements of EISA will gradually increase the average ethanol content of the U.S. gasoline supply (including high-ethanol blends for flex-fuel vehicles) to well above 10 percent. Using illustrative assumptions about the extent to which fuel suppliers would comply with the requirement for advanced biofuels by using biomass-based diesel, CBO estimates that full compliance with the EISA mandates could require the average ethanol content of blended gasoline to reach about 25 percent by 2022 (see Figure 5). For retail gasoline markets to accommodate that much ethanol—while limiting the ethanol content of the blended gasoline that most drivers use to 10 percent—a very large increase in the use of high-ethanol blends would be necessary.

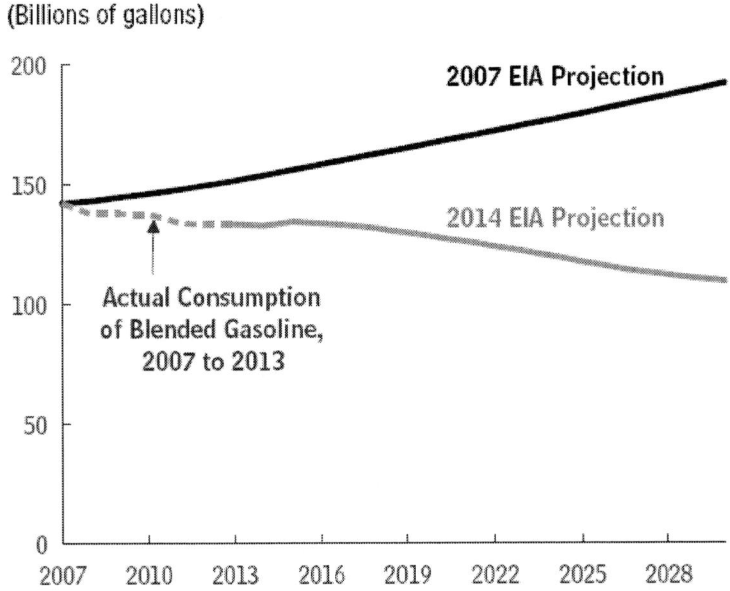

Source: Congressional Budget Office based on Energy Information Administration, *Annual Energy Outlook 2007, With Projections to 2030,* DOE/EIA-0383(2007) (February 2007), www.eia.gov/oiaf/archive/aeo07, and *Annual Energy Outlook 2014, With Projections to 2040,* DOE/EIA0383(2014) (April 2014), www.eia.gov/forecasts/aeo.

Note: EIA = Energy Information Administration.

Figure 4. Changing Expectations About the Future Consumption of Blended Gasoline.

One possibility for raising the total amount of ethanol that the market can accommodate is to boost both the number of flex-fuel vehicles on the road and the extent to which drivers of those vehicles refuel with E85 rather than with conventional blends, such as E10. Flex-fuel technology is relatively inexpensive—adding a few hundred dollars to the manufacturing cost of a new vehicle— so consumers would not need large incentives to buy more of those vehicles. However, substantially increasing the use of E85 would also require increasing the number of filling stations that offer such fuel. Fewer than 2 percent of stations in the United States currently sell E85, although the number has been rising steadily in recent years (it grew fivefold between 2005 and 2012, to more than 2,500 stations).[6] Another factor limiting sales of E85 is its price: Although E85 costs less than regular E10 gasoline, it also has a lower energy content, meaning that it offers fewer miles per gallon. Drivers who could use E85 would be willing to buy it only if its price was low enough relative to the price of E10 to compensate for its lower energy content and potentially for the need to drive farther to find an E85 fueling station.

Although consumption of E85 has been expanding rapidly—by more than 20 percent a year in many recent years—it still accounts for only a tiny fraction of the fuel that passenger vehicles use. If its recent growth rate continues, annual consumption of E85 will reach just 1 billion gallons by 2022, out of a total of 125 billion gallons of blended gasoline projected to be used in that year.

Another possibility for raising the average concentration of ethanol in the fuel supply above 10 percent is to make blended gasoline with up to 15 percent ethanol content (E15) widely available. EPA has certified that vehicles built since 2001—roughly 60 percent of vehicles now on the road—can run on E15 without risking corrosion damage to their fuel lines and engine parts. Many auto-makers disagree and have discouraged their customers from using E15.[7] However, some major manufacturers— including Ford and General Motors—have stated that their models from 2012 or 2013 and later can use E15 without risk.

Experience with vehicles running on E15 has been limited because, until mid-2012, no filling stations offered that fuel. In recent years, the Department of Agriculture provided funding (through the Rural Energy for America Program) for installing pumps that can dispense either E10 or E15; currently, a small number of stations have E15 pumps.[8] But because filling stations that would like to offer both blends would incur costs to acquire new pumps and underground storage tanks, the growth of E15 sales is expected to be slow. In addition, some station owners may be concerned about potential liability claims arising from drivers who inadvertently refuel a pre-2001 vehicle with E15.

A final possibility for addressing the blend wall is to rely more on "drop-in" fuels made from cellulose. The same sorts of cellulosic feedstocks that are used to make biofuels can also be used to produce gasoline or diesel. Those drop-in fuels are identical to conventionally made gasoline and diesel and can substitute for them in full, rather than having to be blended into conventional fuel. The technologies for making any kind of cellulosic fuel are new, however, and production remains costly. (In addition, only a small fraction of the cellulosic production plants projected to open in the next few years are expected to make drop-in fuels.) Nevertheless, to the extent that production of cellulosic gasoline and diesel grows, using more of those drop-in fuels can increase the renewable content of the nation's supply of transportation fuel without exacerbating concerns about the blend wall.

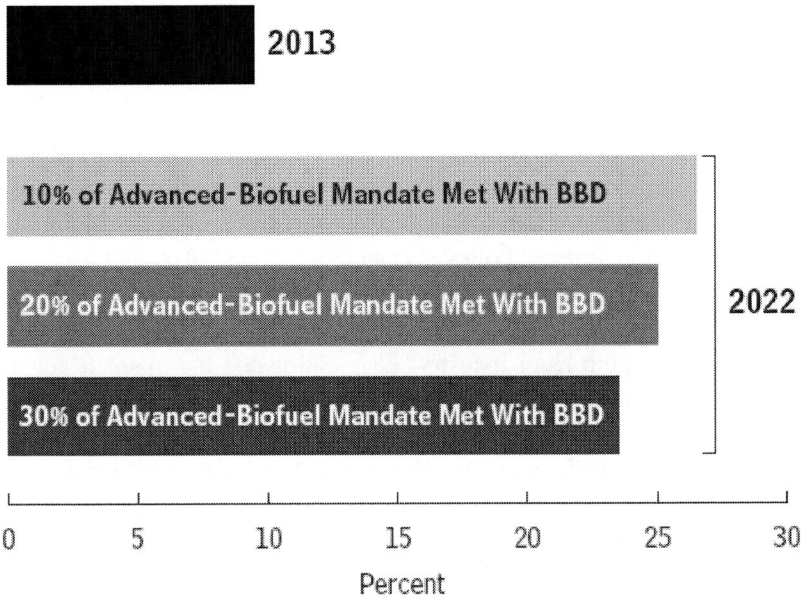

Source: Congressional Budget Office based on Energy Information Administration, Annual Energy Outlook 2014, With Projections to 2040, DOE/EIA-0383(2014) (April 2014), www.eia.gov/forecasts/aeo.

Notes: CBO's calculations are based on the Energy Information Administration's most recent projection of the use of blended gasoline. CBO's estimate of the percentage of ethanol in blended gasoline depends on how much biomass-based diesel is used to comply with the mandate for advanced biofuels. In evaluating the effects of different amounts of use, CBO assumed that the total energy consumption of blended gasoline would remain the same.

BBD = biomass-based diesel.

Figure 5. Ethanol as a Percentage of Blended Gasoline Under Different Assumptions About the Future Use of Biomass-Based Diesel.

EPA's Response to Compliance Challenges

The Energy Independence and Security Act requires that EPA evaluate the Renewable Fuel Standard's requirements each year and adjust them, if necessary, on the basis of market conditions. EPA's response to the gap between the RFS mandate on cellulosic biofuels and actual production of those fuels has been to use its waiver authority to significantly alter that mandate.

For 2010, the first year the cellulosic biofuel mandate was in effect, EPA reduced the requirement of 100 million gallons stated in EISA to 6.5 million gallons—the amount that fuel suppliers could comply with using RINs they had obtained in previous years by exceeding those years' requirements. (The earlier requirements were based on a broader definition of cellulosic biofuels, as described in the Energy Policy Act of 2005.) For 2011 and 2012, EPA initially reduced the cellulosic biofuel mandates significantly. However, following negligible production of cellulosic biofuels in those years and court challenges by the petroleum industry, EPA eliminated the mandate for 2012 and has proposed doing the same for 2011.[9] In addition, the agency lowered the 2013 requirement from 1 billion gallons to less than 1 million gallons (reflecting the industry's production capacity in that year).[10] EPA did not also reduce the requirements for total renewable fuels or for advanced biofuels when it lowered those cellulosic mandates; fuel suppliers were able to make up for the lack of cellulosic biofuels mainly by using biomass-based diesel and noncellulosic ethanol made from sugarcane.[11]

For 2014, EPA has proposed reducing the cellulosic biofuel requirement from 1.75 billion gallons to 17 million gallons.[12] For the first time, it has also proposed decreasing the RFS mandates on total advanced biofuels and total renewable fuels: Those requirements would shrink by more than 1.5 billion gallons and by nearly 3 billion gallons, respectively (from 3.75 billion to 2.2 billion gallons of advanced biofuels and from 18.15 billion to 15.21 billion gallons of renewable fuels). EPA's proposal reflects concern that the 2014 mandate on total renewable fuels in EISA would cause the average ethanol content of the nation's gasoline supply to exceed the 10 percent concentration that many non-flex-fuel vehicles can use. To maintain a proportional cap on the use of corn ethanol, EPA has also proposed reducing by more than 1 billion gallons the portion of the RFS that does not have to be met with advanced biofuels (from 14.4 billion to 13.01 billion gallons).

The annual mandates for cellulosic biofuels specified in EISA through 2022 are so much greater than the industry's projected capacity that EPA will probably continue to reduce the mandate every year, rather than impose large fines on fuel suppliers that are unable to meet the requirement because the fuels are not available. However, granting fuel suppliers a waiver for cellulosic biofuels is likely to have the unintended effect of slowing the growth of production capacity for such fuels by weakening incentives for the private sector to invest in building that capacity. Similar effects would occur for other advanced biofuels if the mandates for those fuels were reduced. In addition, if EPA continues to lower the annual requirements for total renewable fuels to avoid exceeding the blend wall, it will lessen incentives to expand the number of filling stations that offer E85, even though such expansion would help retail gasoline markets accommodate more ethanol in the fuel supply.

EFFECTS OF THE RFS ON THE USE OF RENEWABLE FUELS

Until this year, fuel suppliers were required to use the total volume of renewable fuels stated in EISA and to comply with that law's cap on the use of corn ethanol. Looking ahead, EPA or the Congress may set the total requirement for renewable fuels, the nested mandates (such as the requirement for advanced biofuels), and the corn ethanol cap at levels below those specified in EISA, as EPA has proposed doing for 2014.

To illustrate how the Renewable Fuel Standard—and potential changes to it—might affect the use of renewable fuels over the next several years, CBO estimated the amount of renewable fuels that would be consumed in 2017 under three alternative scenarios: if fuel suppliers had to comply with the requirements stated in EISA (other than the cellulosic biofuel mandate), if those requirements were set at the amounts currently proposed for 2014, and if lawmakers immediately repealed the RFS.

EISA Volumes Scenario

The first scenario represents what would be likely to occur if EPA did not alter the total requirement for renewable fuels, the advanced-biofuel mandate, the biomass-based diesel mandate, and the corn ethanol cap specified in EISA—for example, if the courts or lawmakers prevented EPA from making such modifications.

Requirements

Under the EISA volumes scenario, fuel suppliers would be required to use the following in 2017 (see Table 1):

- 24 billion gallons of renewable fuels in all, including
- 9 billion gallons of advanced biofuels, of which roughly 2 billion compliance-equivalent gallons would have to be biomass-based diesel,[13] and
- No more than 15 billion gallons of corn ethanol.

Table 1. Use of Renewable Fuels in 2017 Under CBO's Alternative Scenarios for the Renewable Fuel Standard

	EISA Volumes Scenario[a]		2014 Volumes Scenario[b]		Repeal Scenario[c]
	Volume Requirement (Billions of gallons)	Blend Requirement Percent)[d]	(Billions of gallons) Volume Requirement	Blend Requirement (Percent)[d]	Estimated Volume (Billions of gallons)
Advanced Biofuels					
Biomass-based diesel[e]	2.0	1.3	1.9	1.2	Less than 1
Other advanced biofuels	7.0	4.0	0.3	0.1	[f]
Subtotal	9.0	5.3	2.2	1.3	Less than 1
Corn Ethanol	15.0	[g]	13.0	[g]	13[h]
Total Renewable Fuels	24.0	14.5	15.2	9.2	13 to 14

Source: Congressional Budget Office based on section 202 of the Energy Independence and Security Act of 2007; Environmental Protection Agency, *EPA Proposes 2014 Renewable Fuel Standards, 2015 Biomass-Based Diesel Volume,* Regulatory Announcement EPA-420-F13-048 (November 2013), http://go.usa.gov/9rdA (PDF, 190 KB); and Energy Information Administration, "Annual Energy Outlook 2014: Market Trends—Oil/Liquids" (April 2014), www.eia.gov/forecasts/aeo/MT_liquidfuels.cfm.

[a] For this scenario, CBO assumed that fuel suppliers would have to comply with the total requirement for renewable fuels and the cap on corn ethanol that are specified for 2017 in the Energy Independence and Security Act of 2007 (EISA). Those requirements mean that fuel suppliers would also be required to use 9

billion gallons of advanced biofuels, with specific quantities consisting of biomass-based diesel and cellulosic biofuels. The Environmental Protection Agency (EPA) has not yet specified the requirement for biomass-based diesel for 2017 (EISA mandates that it be at least 1.5 billion gallons, measured in compliance-equivalent gallons). For illustrative purposes, CBO assumed that fuel suppliers would be required to use 1.33 billion gallons of biomass-based diesel (which count as 2.0 billion compliance-equivalent gallons)—the amount of biomass-based diesel that the Energy Information Administration projects will be used in 2017. The 7 billion gallons of advanced biofuels not composed of biomass-based diesel would include a minimum quantity of cellulosic biofuels, which has not yet been specified by EPA.

[b] For this scenario, CBO assumed that the 2017 requirements for renewable fuels would be set at the same volumes that EPA has proposed for 2014. Thus, the 0.3 billion gallons of other advanced biofuels would have to include at least 17 million gallons of cellulosic biofuels. Total use of transportation fuels in the United States is projected to be similar in 2017 and 2014, so this scenario would make the Renewable Fuel Standard about as stringent in 2017 as it is in 2014.

[c] For this scenario, CBO assumed that lawmakers would repeal the Renewable Fuel Standard in 2014, so fuel suppliers would not be subject to any requirements for the use of renewable fuels in 2017.

[d] EPA translates the annual volume requirements in EISA into percentage blend requirements using projections of the total amount of gasoline and diesel that will be used in a given year. Those requirements specify the percentages of various renewable fuels that suppliers must blend into gasoline or diesel to comply with the EISA mandates. CBO estimated the percentage requirements for 2017 using the relationship between the volume requirements and blend requirements that EPA calculated for 2014 (because total U.S. consumption of gasoline and diesel is projected to be similar in those two years).

[e] Figures for biomass-based diesel are measured in compliance-equivalent gallons. Under EISA, 1 gallon of biomass-based diesel is considered equivalent to 1.5 gallons of other types of advanced biofuels or of corn ethanol for the purposes of complying with the Renewable Fuel Standard. EPA's proposed standard for 2014 is 1.28 billion gallons of biomass-based diesel, which would equal 1.92 billion compliance-equivalent gallons.

[f] If lawmakers repealed the Renewable Fuel Standard, fuel suppliers would probably continue to use small quantities of other advanced biofuels in addition to biomass-based diesel. Those quantities would include sugarcane ethanol used to meet state requirements for renewable fuel use as well as the small amounts of cellulosic biofuels that would continue to be produced at existing plants.

[g] The volume specified for corn ethanol is an upper limit on its use rather than a minimum requirement, so EPA does not calculate a percentage blend requirement for corn ethanol.

[h]This figure is based on the expectation that corn ethanol will make up roughly 10 percent of the 132 billion gallons of blended gasoline projected to be used in the United States in 2017.

In addition to those requirements, EISA mandates that 5.5 billion gallons of the advanced-biofuel requirement be met using cellulosic biofuels. For this scenario, CBO assumed that EPA would continue to reduce the requirement for cellulosic biofuels to the amount that could be made from available production capacity—projected by the Energy Information Administration to be about 170 million gallons in 2017—and that fuel suppliers would be allowed to use other types of advanced biofuels to make up the remaining volume for the advanced-biofuel mandate.

Effects on Use

The requirements of EISA outlined above imply that fuel suppliers would have to use about 7 billion gallons of advanced biofuels of some sort in 2017 in addition to roughly 2 billion gallons of biomass-based diesel. What types of fuel they would use to meet that goal is highly uncertain. To date, no more than 500 million additional gallons of advanced biofuels have been required (beyond the mandate for biomass-based diesel).[14] Suppliers have met that additional quantity by using slightly more biomass-based diesel than required and by importing sugarcane ethanol. Increasing the use of those types of advanced biofuels enough to meet the 7 billion gallon requirement—especially over just a few years—would probably be challenging and costly. For example, consider the following illustrative increases in advanced biofuels:

- The Energy Information Administration currently projects that the United States will use roughly 2 billion gallons of biomass-based diesel (measured in compliance-equivalent gallons) in 2017 and expects annual consumption to remain constant through 2040.[15] To use an additional 2 billion compliance-equivalent gallons of biomass-based diesel would mean doubling the projected supply of that fuel, which would most likely require a significant increase in its price.
- To import an additional 4 billion gallons of sugarcane ethanol from Brazil (the primary source for the sugarcane ethanol used in the United States) would require a 45 percent increase in Brazil's production from the amount projected for 2017 [16] Fostering such a large increase in production in a short time would be difficult—and would probably require a significant increase in the price of sugarcane

ethanol—given the time lags involved in planting and harvesting a perennial crop such as sugarcane and the need for additional production capacity and transportation infrastructure [17]

Even such large boosts in supply—coupled with the projected 170 million gallons of production capacity for cellulosic biofuels—would leave about 1 billion gallons of the 7 billion gallon gap between EISA's mandates for advanced biofuels and biomass-based diesel in 2017 to be filled by other advanced biofuels. Rising prices for advanced biofuels could encourage the production of new supplies based on additional feedstocks, such as sorghum. However, at present, little information exists to project how large such supplies would be and how they could become available by 2017.

2014 Volumes Scenario

In the second scenario, CBO assumed that the requirements for various types of renewable fuels in the next few years would be set at the same volumes that EPA has proposed for 2014. Total U.S. consumption of transportation fuels is projected to be similar in 2017 and 2014, so this scenario would make the Renewable Fuel Standard about as stringent then as it is now.

In the absence of Congressional action or legal restrictions, CBO considers this scenario much more likely than the EISA volumes scenario, which would require a large and rapid increase in the use of advanced biofuels and would cause the total percentage of ethanol in the gasoline supply to rise to levels that would require significant changes in the infrastructure of fueling stations.

Requirements

Under the 2014 volumes scenario, fuel suppliers would be required to use the following in 2017:

- 15.2 billion gallons of renewable fuels in all, including
- 2.2 billion gallons of advanced biofuels, of which 1.9 billion compliance-equivalent gallons would have to be biomass-based diesel, and
- No more than about 13 billion gallons of corn ethanol.

As in the EISA volumes scenario, CBO assumed that EPA would reduce the mandate on cellulosic biofuels for 2017 to a level consistent with the production capacity currently projected for that year (about 170 million gallons).

Effects on Use

CBO expects that fuel suppliers would just meet those requirements but not exceed them.

Repeal Scenario

The final scenario represents CBO's assessment of what would happen if lawmakers immediately eliminated the RFS, and fuel suppliers used renewable fuels only to the extent that doing so was cost-effective for them.

Requirements

Under the repeal scenario, fuel suppliers would have no requirements to use specific types or amounts of renewable fuels.

Effects on Use

CBO estimates that in the absence of the Renewable Fuel Standard, fuel suppliers would use less than 1 billion gallons of biomass-based diesel in 2017 and roughly 13 billion gallons of corn ethanol. That amount of biomass-based diesel is less than half the consumption (measured in compliance-equivalent gallons) that is projected for 2014, including the effects of the RFS. That decline would occur, in CBO's estimation, because about half of all biomass-based diesel is made from soybean oil, and available evidence suggests that the cost of producing diesel from soybean oil is higher than the wholesale price of petroleum-based diesel.[18] Biomass-based diesel is also made from several other materials, such as recycled vegetable oil or animal fat. Some of those types of biomass-based diesel (particularly diesel made from waste products) would probably remain cost-effective even without the incentives created by the RFS because the production materials are generally available at a relatively low cost.[19]

Consumption of corn ethanol would be about the same under the repeal scenario as under the 2014 volumes scenario, CBO estimates. Because ethanol is expected to cost less per gallon than gasoline in 2017, fuel suppliers would probably find it profitable to use 13 billion gallons of ethanol in that year—the volume that corresponds to the maximum blend of ethanol in gasoline (10 percent) that virtually all vehicles now on the road can use.[20] Even if ethanol did not have a price advantage, it would probably continue to be in demand to some extent because of its other benefits. In particular, adding ethanol helps suppliers ensure that their fuel meets emission limits for carbon monoxide (an air pollutant regulated by EPA) and octane requirements (for improved vehicle performance).

Over the longer term, the effect of a repeal on the use of ethanol could be greater. For example, the per-gallon price of corn ethanol might rise above that of gasoline, causing fuel suppliers to reduce the concentration of ethanol in gasoline below current levels. Another possibility is that future advances in technology could allow the development of cost-effective octane-enhancing substitutes for ethanol, which could cause fuel blenders to favor the use of those substitutes. If so, ethanol consumption under the repeal scenario could fall short of that under the other scenarios by growing amounts.

Effects of the RFS on Prices and Spending for Food

To the extent that the Renewable Fuel Standard raises the demand for ethanol made from cornstarch, it will increase corn prices and thus prices for the wide variety of foods that are produced with corn—ranging from corn syrup sweeteners to meat, dairy, and poultry products.[21] Some policymakers have expressed concern about the size of those potential price increases and their effects on households' food spending. Although food prices depend on many uncertain factors, CBO's analysis suggests that differences in food prices and spending under the agency's three scenarios for the RFS would probably be small. Specifically, CBO estimated that total U.S. food expenditures in 2017 would be higher under the EISA volumes scenario than under the other two scenarios by $3.5 billion, or about 0.2 percent of the approximately $1.8 trillion in spending on food expected in 2017.[22] The basis for that estimate is described below (the calculations are discussed in detail in the appendix).

How the RFS Affects the Use of Corn Ethanol

A key consideration when evaluating the effect of the renewable fuel mandates on food prices is the extent to which the use of corn ethanol differs among the scenarios. As described above, CBO expects that consumption of corn ethanol would be about the same whether the RFS was repealed or the 2017 requirements were set at the proposed 2014 volume of 13 billion gallons. If, by contrast, fuel suppliers had to meet the 2017 requirements specified in EISA, corn ethanol use would total 15 billion gallons in that year, CBO estimates. That additional 2 billion gallons would raise the total demand for corn.

How the Demand for Corn Ethanol Affects the Price of Corn

Of the U.S. corn supply, roughly 15 percent is used for food products, 40 percent for animal feed, and 40 percent for ethanol production (the rest is exported). Thus, any significant change in the demand for corn ethanol that resulted from the RFS could have a noticeable effect on corn prices. The extent to which corn prices would be affected would depend on how sensitive the supply of and demand for corn are to changes in its price. Analysts have produced a range of estimates for that sensitivity (known as an elasticity). Using estimates that are in the middle of that range, CBO projects that consumption of an extra 2 billion gallons of corn ethanol under the EISA volumes scenario would increase corn prices in 2017 by about 25 cents per bushel (roughly 6 percent) compared with prices under the other two scenarios. That estimate takes into account the extent to which higher prices would boost corn production and reduce nonethanol uses of corn (such as for food or animal feed), both of which would limit some of the price increase that would otherwise result.

The difference in corn prices between the scenarios could be larger over the longer term. If, following the repeal of the RFS, the ethanol content of the gasoline supply fell below 10 percent, the gap between ethanol use under that standard and in its absence would widen. As a result, differences in the consumption of corn ethanol, and thus in the price of corn, between the EISA volumes scenario and the repeal scenario would grow over time.

How the Price of Corn Affects the Cost of Food

Higher corn prices would increase food prices directly because of the large variety of food products that contain corn. Higher corn prices would also operate indirectly through two different mechanisms. First, higher prices for corn used as animal feed would lead to price increases for meat, poultry, and dairy products. Second, higher corn prices would cause farmers to produce corn in place of other crops, such as soybeans, and decreased production of those crops would in turn raise their prices.

With both direct and indirect effects included, full compliance with the EISA mandates would increase total spending on food in 2017 by $3.5 billion, CBO estimates, relative to spending under the other two scenarios. With total U.S. food expenditures expected to be roughly $1.8 trillion in 2017, that increase would represent a rise of about 0.2 percent.

Changes in food prices could affect federal programs that are linked to those prices, such as the Supplemental Nutrition Assistance Program, or SNAP (formerly known as Food Stamps), and various programs that provide meals to children at school and in other settings. Once a year, the government adjusts the benefits paid under SNAP and the child nutrition programs on the basis of shifts in food prices. As a result, changes in food prices would lead to roughly proportionate changes in spending on such benefits.

EFFECTS OF THE RFS ON PRICES OF TRANSPORTATION FUELS

The Renewable Fuel Standard boosts the use of renewable fuels by requiring fuel suppliers to obtain a specific number of RINs (with each RIN corresponding to a gallon of renewable fuel that has been blended into the fuel supply) for every gallon of petroleum-based gasoline or diesel that they use. How that requirement affects the prices of various fuels depends on a fuel's composition of petroleum-based and renewable elements. To better understand the potential size of those effects over the next several years, CBO estimated how the price of diesel and E10—the two most commonly consumed transportation fuels—and the price of E85 would differ in 2017 among its three scenarios for the RFS.

Effects under the 2014 Volumes Scenario or the Repeal Scenario

CBO concludes that prices for diesel, E10, and E85 would probably be essentially the same in 2017 whether the RFS requirements were kept at the amounts proposed for 2014 or the RFS was repealed. As explained above, fuel suppliers would probably use roughly 13 billion gallons of corn ethanol (corresponding to a 10 percent blend) under the repeal scenario—the same amount that they would be required to use under the 2014 volumes scenario. The main difference in consumption of renewable fuels in 2017 between those two scenarios is that the 2014 volumes scenario would require the use of up to 2 billion more gallons of biomass-based diesel than fuel suppliers would use under the repeal scenario. However, that additional use would probably not have a significant effect on the prices of transportation fuels because the increase would be small relative to the total volume of fuel.

Effects under the EISA Volumes Scenario

If EPA set the total requirement for renewable fuels and the cap on corn ethanol at the 2017 volumes stated in EISA, fuel suppliers would have to use up to 11 billion gallons more of renewable fuels than they would choose to use under the repeal scenario (see Table 1 on page 11). Those extra gallons would consist of the following:

- 2 billion additional gallons of corn ethanol; and
- Up to 9 billion additional gallons of advanced biofuels, which could be made up of biomass-based diesel, sugarcane ethanol, or other qualifying fuels.

Prices for diesel and E10 would probably be higher under the EISA volumes scenario than in the absence of the RFS, whereas the price of E85 would probably be significantly lower.[23] CBO reached those conclusions by examining the requirements that suppliers of those fuels would face and the costs of meeting the requirements.

Compliance Requirements

To implement the Renewable Fuel Standard, EPA translates the volume requirements stated in EISA into percentage blend requirements, which equal the mandated volume of each category of renewable fuel divided by the projected volume of gasoline and diesel that is subject to EISA. Those percentage obligations are applied to each fuel supplier's actual sales of gasoline and diesel to determine the number of RINs that the supplier must submit. Fuel suppliers obtain RINs by purchasing qualifying gallons of renewable fuels and blending them into the fuel they sell or by purchasing RINs from suppliers that have accumulated excess RINs by using more renewable fuel than the RFS requires.

Given the percentage blend requirements of the EISA volumes scenario (shown in Table 1 on page 11), for each 100 gallons of diesel or gasoline that a fuel supplier used in 2017, it would need to submit 14.5 RINs to EPA, of which 5.3 would have to qualify as advanced biofuels. Of those 5.3 advanced-biofuel RINs, at least 1.3 would have to be biomass-based diesel RINs. Thus, taking into account the nested nature of the standard, for each 100 gallons of diesel or gasoline it used, a fuel supplier would have to submit the following to EPA:

- 1.3 biomass-based diesel RINs,
- 4.0 advanced-biofuel RINs (the total of 5.3 advancedbiofuel RINs minus the 1.3 biomass-based diesel RINs), and
- 9.2 renewable fuel RINs (the total of 14.5 renewable fuel RINs minus the 5.3 total of advanced-biofuel RINs).

Estimating the effects of the EISA volumes scenario on the prices of diesel, E10, and E85 requires estimating how the Renewable Fuel Standard would affect the price of each type of RIN. It also involves calculating RIN requirements on the basis of the percentages of petroleum-based and renewable fuels in the fuel that a supplier sells.

Effect on the Price of a Renewable Fuel RIN

In past years, fuel suppliers have been able to comply with the RFS (other than the mandate on biomass-based diesel) largely by mixing corn ethanol into gasoline and selling that blend as E10. As a result, renewable fuel RINs have been plentiful, and their price has been low (less than 10 cents in 2011 and 2012).[24] If fuel suppliers had to comply with the EISA volumes scenario, however, the amount of ethanol that would need to be blended into the fuel supply in 2017 would significantly exceed the amount that could be sold as E10. As a result, the price of renewable fuel RINs would rise as suppliers sought to buy more RINs than could be generated through sales of E10. Higher RIN prices would create an incentive for fuel suppliers to sell high-ethanol blends, such as E85, which would yield excess RINs that they could sell. CBO expects that E85 suppliers would use some of the profits they obtained by selling those excess RINs to cover the capital costs necessary to modify more stations to carry E85 (estimated at roughly 10 cents per gallon of E85) and would pass the remaining profits on to consumers in the form of lower E85 prices to encourage sales of that fuel.[25]

With consumption of blended gasoline projected to total 132 billion gallons in 2017, the amount of ethanol that would be blended into the fuel supply under the EISA volumes scenario would exceed the amount used in E10 (13 billion gallons) by at least 6 billion gallons, under CBO's illustrative assumption that 4 billion gallons of the advanced-biofuel requirement would be met using sugarcane ethanol. The remaining 2 billion of those additional 6 billion gallons of ethanol would be corn ethanol. The total amount of additional ethanol used could be greater than 6 billion gallons because CBO assumed the use of 1 billion gallons of unspecified advanced biofuels, which could include ethanol made from other feedstocks, such as sorghum.

For the gasoline supply to absorb an additional 6 billion gallons of ethanol—without exceeding a 10 percent blend in the most commonly used fuel—the use of E85 by flex-fuel vehicles would need to be roughly seven times greater in 2017 than the amount anticipated for 2014. Such an increase would require a large expansion in the number of fueling stations equipped to sell E85. It would also mean that the drivers of the roughly 17 million flex-fuel vehicles expected to be on the road in 2017 would have to refuel almost exclusively with E85—which would require them to refuel more often (because of the lower energy content of E85) and possibly to drive much farther to obtain E85.[26]

Using data from Brazil, where E85 fueling stations are plentiful, researchers estimated that the fuel cost of driving a mile on E85 had to be 15 percent less than the fuel cost of driving a mile on E10 to induce 80 percent of Brazilian owners of flex-fuel vehicles to use E85.[27] However, given the limited availability of E85 pumps in the United States and the unprecedented expansion in the use of E85 that would be necessary to absorb 6 billion additional gallons of ethanol, CBO estimates that savings on fuel costs for E85 drivers in the United States would have to be more substantial—probably about 30 percent to 50 percent lower than the fuel cost of driving with E10.[28] Moreover, filling stations would need to install additional E85 pumps. CBO estimates that a price of roughly $1.45 to $2.00 for renewable fuel RINs would be necessary to cover the average capital costs that filling station owners would incur to install more pumps (about 10 cents per gallon of E85 sold) and to subsidize the price of E85 enough to make the cost of driving a mile with that fuel about 30 percent to 50 percent lower than the cost of driving a mile with E10.[29]

In addition to those subsidy costs, the price of a renewable fuel RIN for a gallon of corn ethanol would have to include an increase of 10 cents per gallon in the price of corn ethanol—the increase that CBO estimates would be necessary to bring about the additional 2 billion gallons of corn ethanol required under the EISA volumes scenario.[30] With all of those price components combined, supplying and using those extra 2 billion gallons of corn ethanol would necessitate a price range of $1.55 to $2.10 for renewable fuel RINs.

Effect on the Price of an Advanced-Biofuel RIN

The EISA volumes scenario would require fuel suppliers to use up to 9 billion more gallons of advanced biofuels than they would in the absence of the RFS, CBO estimates (see Table 1 on page 11). As noted above, those additional gallons would probably consist mainly of biomass-based diesel and sugarcane ethanol, although other advanced biofuels, such as sorghum ethanol, could make up some of the supply. For this analysis, CBO assumed that 4 billion gallons of sugarcane ethanol and 2 billion gallons of biomass-based diesel would be used to help meet the advanced-biofuel mandate (in addition to the 2 billion gallons used to comply with the mandate for biomass-based diesel).

If, as CBO assumes, fuel suppliers did rely on advanced forms of ethanol (such as sugarcane ethanol) to meet 4 billion gallons of the advanced-biofuel requirement, the price of an advanced-biofuel RIN would include the aforementioned $1.45-to-$2.00 subsidy necessary to increase sales of E85 enough to absorb the additional 6 billion gallons of ethanol used under the EISA volumes scenario. However, fuel suppliers would avoid the need to subsidize the use of high-ethanol blends to the extent that they used biomass-based diesel to meet the advancedbiofuel requirement. Thus, the price that suppliers would be willing to pay for RINs generated from biomass-based diesel would include the value of that avoided cost. More generally, as long as fuel suppliers complied with the RFS by using more biomass-based diesel RINs than those necessary to meet the biomass-based diesel mandate (with the additional ones being used to meet the advancedbiofuel mandate), biomass-based diesel RINs would sell for the same price as advanced-biofuel RINs.

Besides the E85 subsidy value, the second component of the price of an advanced-biofuel RIN would be the price premium necessary to cover producers' incremental costs of increasing the supply of advanced biofuels. Complying with the EISA volumes scenario would require a very large increase in the supply of advanced biofuels, so that price premium would probably be large. To expand the supply of biomass-based diesel 2 billion gallons beyond the amount needed to satisfy the mandate for that fuel would require doubling U.S. production of biomass-based diesel. Likewise, importing 4 billion gallons of sugarcane ethanol would require a 45 percent increase in Brazil's production of such ethanol (if Brazil's own consumption did not change) and a more than fourfold increase in the country's exports of sugarcane ethanol from the 2013 level.[31] Even with those increases, complying with the EISA volumes scenario would still require an additional 1 billion gallons of advanced biofuels. (A small fraction of that requirement would be met using cellulosic biofuels, whose production capacity is projected to be 170 million gallons in 2017. CBO assumes that the remaining gallons would be of an unspecified type of advanced biofuel, supplied at the same price as the sugarcane ethanol and the additional biomass-based biodiesel.)

For illustrative purposes, CBO considered the implications for fuel prices if the price premium necessary to induce an additional 7 billion gallon supply of advanced biofuels under the EISA volumes scenario was in the range of \$1.50 to \$4.00 per gallon. The lower end of that range is 50 cents higher than the average spot price (roughly \$1) of an advanced-biofuel RIN or biomass-based diesel RIN in 2011 and 2012. (Because the market did not expect the blend wall to be exceeded in 2011 or 2012, that average RIN price included only the price premium necessary to increase the supply of those fuels to the amounts required by the RFS in those years and did not include any subsidy to increase the consumption of E85.) The upper end of the range was chosen because the price premium that might be needed to induce such large increases in the supply of advanced biofuels in 2017 might be several times higher than that 2011–2012 average. Adding the \$1.50-to-\$4.00 price premium to the \$1.45-to-\$2.00 per-gallon subsidy for E85 suggests that each of the additional 7 billion advanced-biofuel RINs required under the EISA volumes scenario would be valued at roughly \$3.00 to \$6.00.

Effect on the World Price of Oil

Because renewable fuels substitute for gasoline and diesel, they reduce consumption of those fuels in the United States, which could lower the world price of oil and thus the price that fuel suppliers pay for petroleum-based fuels. CBO did not account for that effect in this analysis, for two reasons. First, any such reduction in world oil prices would most likely be small—because, for example, the 10 billion to 11 billion more gallons of renewable fuels consumed under the EISA volumes scenario than under the repeal scenario amount to only about one-half of one percent of the global supply of oil. Second, any policy that tended to reduce U.S. demand for oil, and thus the world price of oil, could be offset if large suppliers of crude oil strategically reduced their production to prevent that price from falling and if consumers worldwide responded to a temporary decline in oil prices by increasing their use of petroleum products.[32]

Effect on the Prices of Transportation Fuels in the United States

The Renewable Fuel Standard affects the prices that consumers pay for various types of fuels differently depending on the mix of petroleum-based and renewable elements used to produce a given fuel. CBO estimated the effects of the EISA volumes scenario on the prices of the two fuels most widely used for transportation in the United States, petroleum-based diesel and E10, as well as on the price of E85. Unless otherwise noted, CBO's calculations reflect the expectation that the prices fuel suppliers pay for RINs, or the profits they receive from selling RINs, will be passed on fully to consumers.

Price of Petroleum-Based Diesel

CBO estimated the effect of the EISA volumes scenario on the price of diesel fuel by applying the range of RIN prices described above to the additional cost components identified for suppliers of diesel. Thus, for each 100 gallons of diesel that a fuel supplier sold, the additional cost that it would incur because of the RFS under the EISA volumes scenario would be the sum of the following:

- 1.3 x the $3.00-to-$6.00 price of a biomass-based diesel RIN,
- 4.0 x the $3.00-to-$6.00 price of an advanced-biofuel RIN, and
- 9.2 x the $1.55-to-$2.10 price of a renewable fuel RIN.

Those costs would add about $30 to $51 for each 100 gallons of petroleum-based diesel, which would raise the average cost of producing diesel in 2017 by 30 cents to 51 cents per gallon, or 9 percent to 14 percent.[33] Price increases would probably be smaller over the longer run, because the substantial increase in RIN prices necessary to cause large changes in renewable fuel use by 2017 would diminish as the market adjusted over time.

Price of E10

For each 100 gallons of E10 that a fuel supplier sells, it uses 90 gallons of petroleum-based gasoline and 10 gallons of corn ethanol. Its RIN requirements are based only on its consumption of gasoline, so those requirements are 10 percent less than if it sold 100 gallons of purely petroleum-based gasoline. In addition to the RIN requirements associated with the 90 gallons of petroleum-based gasoline used, an E10 supplier would pay 10 cents more for each of the 10 gallons of corn ethanol that it used in 2017 under the EISA volumes scenario (the price increase necessary to induce the extra 2 billion gallons of corn ethanol consumed in that scenario). In total, for each 100 gallons of E10 that the supplier sold, the higher cost resulting from the RFS under the EISA volumes scenario would be the sum of the following:

- 0.9 x (1.3 x the $3.00-to-$6.00 price of a biomass-based diesel RIN),
- 0.9 x (4.0 x the $3.00-to-$6.00 price of an advancedbiofuel RIN),
- 0.9 x (9.2 x the $1.55-to-$2.10 price of a renewable fuel RIN), and
- 10 x the $0.10 per gallon increase in the price of corn ethanol.

Those costs would add $28 to $47 for each 100 gallons of E10. However, those higher costs would be partly offset by RINs that the E10 supplier obtained along with each gallon of corn ethanol that it bought. Because the supplier would blend 10 gallons of corn ethanol into every 90 gallons of its fuel supply, it would receive 10 renewable fuel RINs. When that ethanol was blended into the fuel supply, each of those RINs would be worth $1.55 to $2.10.[34] (The E10 supplier would use 8.3, or 0.9 x 9.2, of those renewable fuel RINs to meet its own compliance obligations—offsetting the cost that it would otherwise incur to obtain RINs—and would sell the remaining 1.7.) The value of those 10 RINs ($15.50 to $21) would reduce the additional cost for 100 gallons of E10 to about $13 to $26.[35] Thus, on net, the EISA volumes scenario would raise the average cost of producing a gallon of E10 by 13 cents to 26 cents per gallon in 2017, or 4 percent to 9 percent.[36] As with the price of petroleum-based diesel, the rise in the price of E10 caused by the RFS would probably diminish over time as the market adjusted.

Price of E85

The category of fuel referred to as E85 generally contains between 51 percent and 83 percent ethanol, depending on the season (winter blends have less ethanol to help vehicles start in cold weather). For this analysis, CBO anticipates that E85 will contain an average of 75 percent ethanol and 25 percent gasoline, consistent with recent projections by the Energy Information Administration.[37] Thus, for each 100 gallons of E85 that a fuel supplier sold, it would use 25 gallons of petroleum-based gasoline and 75 gallons of corn ethanol. Its RIN requirements would be based only on its consumption of gasoline, so those requirements would be 75 percent less than if it sold 100 gallons of petroleum-based gasoline. The additional cost that it would incur as a result of the RFS under the EISA volumes scenario would be the sum of the following:

- 0.25 x (1.3 x the $3.00-to-$6.00 price of a biomass-based diesel RIN),
- 0.25 x (4.0 x the $3.00-to-$6.00 price of an advancedbiofuel RIN),
- 0.25 x (9.2 x the $1.55-to-$2.10 price of a renewable fuel RIN),
- 75 x the $0.10 per gallon increase in the price of corn ethanol, and
- 100 x the $0.10 per gallon increase necessary to cover the average capital costs of adding E85 pumps.

Those costs would add $25 to $30 for each 100 gallons of E85, but they would be more than offset by the value of an E85 supplier's renewable fuel RINs. Because the supplier would mix 75 gallons of corn ethanol into every 25 gallons of its fuel supply, it would receive 75 renewable fuel RINs. Once that ethanol was blended into the fuel supply, each RIN would have a value of $1.55 to $2.10. (The supplier would use 2.3, or 0.25 x 9.2, of those RINs to meet its own compliance obligations, leaving 72.7 excess renewable fuel RINs to sell.) The value of those 75 RINs ($116 to $157) would more than offset the fuel suppliers' compliance costs. Thus, on net, the EISA volumes scenario would *decrease* the average cost of a gallon of E85 by 91 cents to $1.27 per gallon in 2017, or 37 percent to 51 percent.[38] The necessary subsidy for E85 would be likely to decrease over time as the availability of E85 pumps grew and the need for drivers to travel out of their way to find E85 declined.

EFFECTS OF THE RFS ON EMISSIONS

EISA sets requirements for how much lower the greenhouse gas emissions associated with renewable fuels must be—relative to the emissions of the gasoline or diesel that they displace—to qualify for use in complying with the RFS. Those requirements vary for different types of renewable fuels:

- Cellulosic biofuels must have emissions that are least 60 percent lower than those of conventional gasoline or diesel fuel.
- Biomass-based diesel and sugarcane ethanol—the main fuels used to comply with the advanced-biofuel mandate—must have emissions that are at least 50 percent lower.
- All other renewable fuels (except those produced at plants operating, or under construction, by December 31, 2007) must have emissions that are at least 20 percent lower. Corn ethanol is the primary renewable fuel in that category.

EPA's estimates of emissions provide the official determination of which biofuels qualify for compliance purposes; however, numerous researchers have estimated the emissions associated with renewable fuels, and their results vary widely. A review of those estimates suggests that total U.S. emissions of greenhouse gases in 2017 would probably be only slightly lower under the EISA volumes scenario than under the other two scenarios described above.

Key Factors That Determine Emissions

Estimating the greenhouse gas emissions associated with the increased use of renewable fuels is complicated, and the results are far from certain. Estimates depend crucially on researchers' projections of many factors, including the following:

- *Crop yields*—In general, higher yields per acre of crops used to produce renewable fuels lead to lower emissions by requiring less land to grow a given amount of feedstocks.
- *Fertilizer use*—Nitrogen, a fertilizer generally used on corn, releases nitrous oxide, a greenhouse gas.
- *Changes in land use and soil carbon*—Carbon is stored in soil and vegetation, and some of it is released as carbon dioxide when previously unfarmed land (such as pasture, grassland, or forest) is brought into production.[39] The RFS can cause the release of stored carbon directly, when crops for renewable fuels are grown on previously unfarmed land, or indirectly, when those crops compete for land that had been used to grow other commodities, leading to higher commodity prices, which encourage growers to put more new land into production. (For example, corn grown for ethanol may displace

corn grown for animal feed or human consumption, leading to higher corn prices and causing previously unfarmed land in the United States or elsewhere to be devoted to growing corn.)

Estimates of the emissions resulting from changes in land use are sometimes large but are also very uncertain. They depend on the type of land brought into production (for instance, carbon dioxide emissions would be much greater if crops were grown on land that had been forest rather than pasture) and, in the case of indirect emissions, on predictions of net changes in land use worldwide. Even changes in crop management practices, such as removing plant residue to produce ethanol, can affect the amount of carbon in the soil.

- *The efficiency of the process for converting feedstocks into fuel*—More-efficient production methods produce more renewable fuels from a given amount of feedstocks and thus generally involve fewer emissions (by requiring less land, fertilizer use, and so forth). Estimates of the efficiency of the production processes for cellulosic biofuels are particularly uncertain because those technologies are still being developed.
- *The generation of electricity credits*—Some methods for producing cellulosic biofuels separate the components of the plants that are converted into sugars, and then into ethanol, from a residual material (called lignin) that can be used to generate electricity. In such cases, most researchers "credit" the production of the biofuel with any reduction in greenhouse gas emissions that occurs when lignin-based generation of electricity is assumed to displace fossil-fuel-based generation.

In addition, because renewable fuels substitute for gasoline, they reduce U.S. consumption of gasoline, which could lead to a slight decline in the world price of oil. Any price decline, in turn, would trigger a "rebound effect"—an increase in oil consumption and emissions— inside and outside the United States.[40] The magnitude of such a rebound effect is very uncertain. It depends on the extent to which the RFS might decrease the global demand for oil, which is dependent on the particular requirements of the RFS and is complicated by uncertainty about whether large suppliers of crude oil (such as the Organization of Petroleum Exporting Countries) would prevent the price of oil from falling by reducing their production.[41] If the RFS did lead to a decline in the world price of oil, that decline would probably be small (for example, the 10 billion to 11 billion additional gallons of renewable fuels used in 2017 under the EISA volumes scenario, compared with the 2014 volumes scenario, equal roughly one-half of one percent of the global supply of oil). Nevertheless, that small price reduction would affect oil consumption around the world.

Researchers' estimates of the rebound effect differ greatly. One recent review of past studies found a range of estimates, suggesting that the rebound effect could offset 29 percent to 85 percent of the reduction in the consumption of crude oil that the RFS would otherwise bring about.[42] Those studies' authors typically assumed a competitive global market for oil; the rebound effect would be smaller if large oil suppliers behaved strategically to limit a price drop.

Most researchers' estimates of the emissions associated with various types of renewable fuels indicate the percentage reduction in emissions that might occur when one gallon of renewable fuel replaced an equivalent amount of gasoline; they do not take any potential rebound effect into account.[43] To the extent that such an effect occurred, it would modify the emission estimates described below—tending to lower estimates of the emission savings, or raise estimates of the emission increases, that additional use of renewable fuels might bring about.

Even without accounting for the rebound effect, researchers produce widely differing estimates of the emission implications of replacing petroleum-based fuels with renewable fuels. Those differences highlight the sensitivity of such estimates to the factors described above.

Emission Estimates for Various Types of Renewable Fuels

Estimates of the greenhouse gas emissions that result from making and using renewable fuels (including emissions created by changes in land use) vary widely according to the feedstocks used to produce the fuel. Much of the research on emission estimates has focused on corn ethanol and cellulosic ethanol, so the discussion below presents more emission estimates for those types of renewable fuels than for others, such as sugarcane ethanol and biomass-based diesel.

Corn Ethanol

EPA estimates that by 2022, the emissions associated with a gallon of corn ethanol would be 21 percent lower than those associated with a gallon of gasoline —just meeting the 20 percent minimum reduction threshold established by EISA.[44] That estimate is calculated assuming that ethanol is produced using the technologies projected to be available in 2022. Accounting for uncertainty in its estimates of the emissions resulting from changes in land use, EPA concludes that there is a 95 percent chance that the actual emission reduction from using corn ethanol in place of gasoline would be between 7 percent and 32 percent (see Figure 6 on page 24). That range is referred to as the 95 percent prediction range.

Critics of EPA's analysis note that its assumption that corn ethanol is produced using the technologies projected to exist in 2022 is likely to overstate the emission reductions caused by the RFS before then.[45] Researchers who do not explicitly assume 2022 technologies have estimated a variety of outcomes. For example, a 2011 report by the National Research Council cites three studies that conclude that substituting corn ethanol for gasoline increases emissions and one study that concludes that such substitution reduces emissions.[46] Another study published that year used simulations to account for the uncertainty in the ethanol production process as well as for the uncertainty associated with changes in land use. It estimated that the emissions resulting from the production and use of corn ethanol were 20 percent *higher* than those resulting from gasoline, with a 95 percent prediction range of 32 percent lower than gasoline to 85 percent higher.[47]

In contrast to that analysis, a 2012 study by another set of researchers estimated that the emissions associated with a gallon of corn ethanol are 34 percent lower than the emissions associated with the gasoline that it displaces (with an 80 percent prediction range of 19 percent to 48 percent lower).[48] The authors stated that their emission estimates were lower than those of other studies because they accounted for recent improvements in corn yields and in the technology used to produce corn ethanol, as well as incorporating improved estimates of emissions resulting from changes in land use.[49]

Sugarcane Ethanol

The sugarcane ethanol used in the United States is generally made in Brazil, and estimates of its emissions include the emissions associated with shipping it to the United States. EPA estimates that sugarcane ethanol emissions would be 61 percent lower than gasoline emissions by 2022, with a 95 percent prediction range of 52 percent to 71 percent lower—exceeding the 50 percent minimum reduction threshold for advanced biofuels. EPA's estimate does not include emissions caused by indirect changes in land use that might be triggered by increased consumption of sugarcane ethanol.[50]

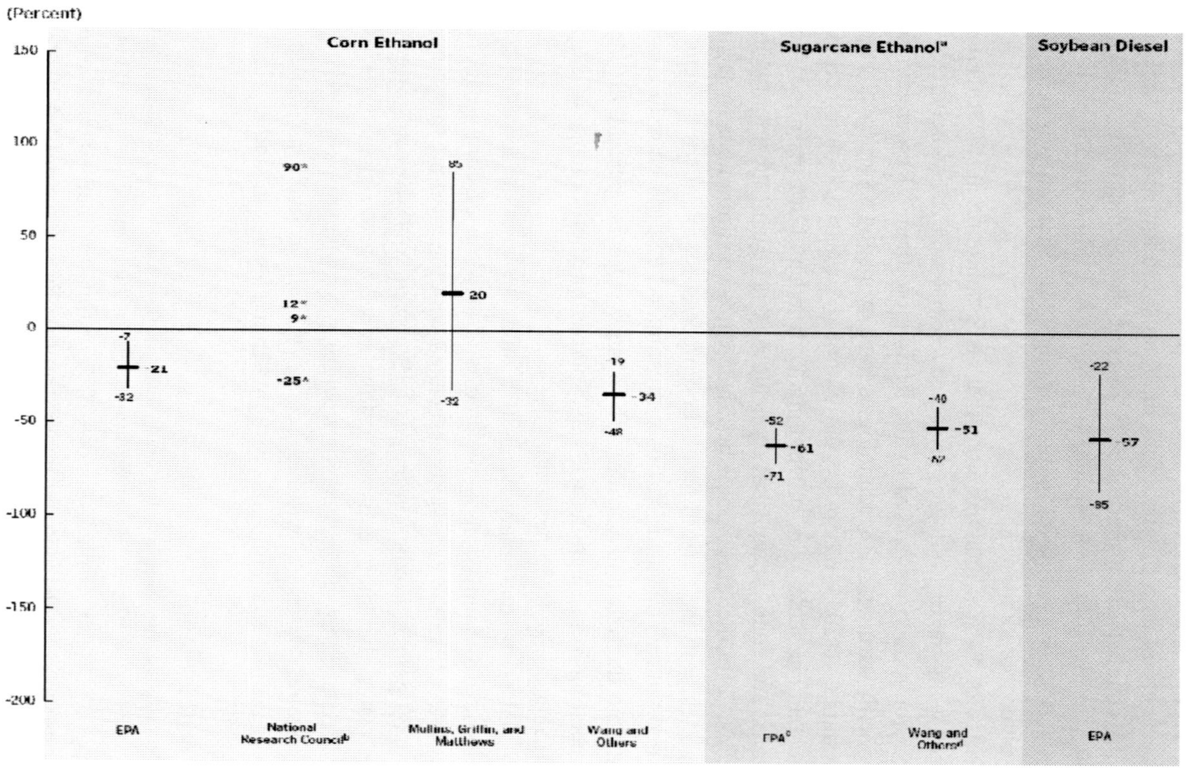

Figure 6. (Continued).

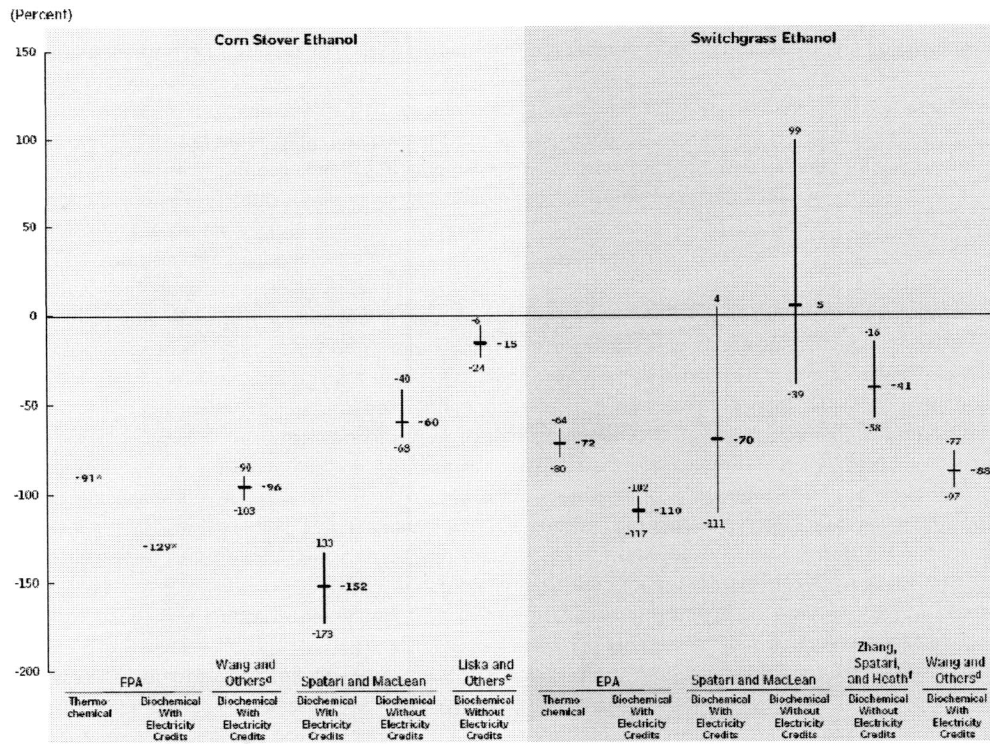

Source: Congressional Budget Office based on Environmental Protection Agency, Renewable Fuel Standard Program (RFS2) Regulatory Impact Analysis, EPA-420-R-10-006 (February 2010), www.epa.gov/otaq/renewablefuels/420r10006.pdf (17 MB); National Research Council, Renewable Fuel Standard: Potential Economic and Environmental Effects of U.S. Biofuel Policy

(National Academies Press, 2011), http://tinyurl.com/89yfx36; Kimberley A. Mullins, W. Michael Griffin, and H. Scott Matthews, "Policy Implications of Uncertainty in Modeled Life-Cycle Greenhouse Gas Emissions of Biofuels," Environmental Science and Technology, vol. 45, no. 1 (January 1, 2011), pp. 132–138, http://dx.doi.org/10.1021/es1024993; Michael Q. Wang and others, "Well-to- Wheels Energy Use and Greenhouse Gas Emissions of Ethanol From Corn, Sugarcane, and Cellulosic Biomass for U.S. Use," Environmental Research Letters, vol. 7, no. 4 (October–December 2012), http://dx.doi.org/10.1088/1748-9326/7/4/045905; Sabrina Spatari and Heather L. MacLean, "Characterizing Model Uncertainties in the Life Cycle of Lignocellulose-Based Ethanol Fuels," Environmental Science and Technology, vol. 44, no. 22 (November 15, 2010), pp. 8773–8780, http://dx.doi.org/10.1021/es102091a; Adam J. Liska and others, "Biofuels From Crop Residue Can Reduce Soil Carbon and Increase CO2 Emissions," Nature Climate Change, vol. 4 (April 20, 2014), http://dx.doi.org/10.1038/nclimate2187; and Yimin Zhang, Sabrina Spatari, and Garvin Heath, "Are We Ready for Consequential LCA-Based Regulations?" (presentation prepared by the National Renewable Energy Laboratory for the American Center for Life Cycle Assessment's LCA X Conference, Portland, Oregon, November 2, 2010), www.lcacenter.org/LCAX/presentations-final/145.pdf (746 KB).

Notes: Unless otherwise noted, the ranges of results shown for individual studies are the 95 percent prediction ranges, with the median estimate indicated. "Electricity credits" denote that the ethanol production process was credited with emission reductions that occurred when components of the feedstocks that could not be used to make ethanol were used to generate electricity, which was then assumed to displace fossil-fuel-fired electricity. Those electricity credits allow emission estimates for cellulosic ethanol (such as corn stover ethanol or switchgrass ethanol) to be more than 100 percent lower than emission estimates for gasoline or diesel. Unless otherwise noted, all estimates include emissions from policy-induced changes in land use.

EPA = Environmental Protection Agency; * = point estimate.

a. Made in Brazil and shipped to the United States.

b. Point estimates for national average outcomes found in studies cited by the National Research Council.

c. Does not include emissions from price-induced changes in land use resulting from the increased demand for sugarcane.

d. Results show the 80 percent prediction range, with the median estimate indicated.

e. Results show the average outcome plus and minus two standard deviations; according to that study, actual emission reductions would be likely to lie within that range.

f. Uses the same model and assumptions as Spatari and MacLean but does not include emissions associated with changes in land use.

Figure 6. Estimated Difference Between the Greenhouse Gas Emissions Associated With Various Biofuels and the Emissions Associated With the Gasoline or Diesel They Replace.

Another set of researchers, who did include an estimate of those emissions, concluded that the emissions associated with sugarcane ethanol were 51 percent lower than those associated with gasoline, with an 80 percent prediction range of 40 percent to 62 percent lower.[51]

Soybean Diesel

About half of the biomass-based diesel consumed in the United States is made from soybean oil. (The next three largest sources of such diesel—corn oil, yellow grease, and alcohol—each account for less than 10 percent of total consumption.)[52] EPA estimates that soybean oil diesel has emissions that are 57 percent lower than those of petroleum-based diesel, with a 95 percent prediction range of 22 percent to 85 percent lower.

Cellulosic Ethanol

The emissions associated with cellulosic biofuels are generally estimated to be low because relatively little energy is needed to grow cellulosic feedstocks and convert them into ethanol (although the conversion is capital-intensive and costly). Moreover, one method for producing cellulosic ethanol—a biochemical process—also results in an organic polymer called lignin, which can be used to generate electricity.

Most emission estimates for ethanol produced using that process credit the ethanol with emission reductions that occur under the assumption that the lignin-generated electricity is sold to the electrical grid, displacing power generated from fossil fuels. Those "electricity credits" allow emission estimates for cellulosic ethanol to be more than 100 percent lower than emission estimates for gasoline or diesel.

Corn Stover

Most researchers conclude that ethanol produced from corn stover—the material left over when corn is harvested—has lower emissions than any other type of cellulosic biofuel. The reason is that corn stover is not expected to trigger any changes in land use (either directly to produce it or indirectly through changes in corn prices) and thus would not generate any land-use-related emissions. The volume of ethanol that can be produced from corn stover is limited by the demand for corn and by the amount of corn stover that can be removed from fields without significantly reducing the productivity of the soil.

EPA predicts (on the basis of technologies expected to be available in 2022) that emissions associated with ethanol made from corn stover would be

91 percent lower than those associated with gasoline if the ethanol was produced using a thermochemical process, or 129 percent lower if it was made using the biochemical process that results in lignin (taking into account electricity credits). Factoring in key uncertainties about the production process, other researchers estimated that emissions resulting from corn stover ethanol (produced using a biochemical process and with electricity credits taken into account) would be 90 percent to 103 percent lower than those from gasoline (based on an 80 percent prediction range).[53] A different set of researchers showed the sensitivity of emission estimates to assumptions about electricity credits.[54] Their median estimate was that the emissions associated with making corn stover ethanol using a biochemical process would be 152 percent lower than the emissions associated with gasoline if credits were included for emission reductions from selling excess electricity to the grid, or 60 percent lower if such credits were not included.

Compared with those estimates, a new study reports a much less favorable outlook for reducing emissions by substituting corn stover ethanol for gasoline. Its different conclusions reflect the authors' estimate that removing corn stover from fields increases emissions because that material releases some of its carbon into the atmosphere at a faster rate when used in biofuel production than when left to decay into the soil. Accounting for the increases in emissions of carbon dioxide that would occur over 10 years, they estimate that, on average, the emissions associated with corn stover ethanol would be 15 percent less than those associated with gasoline, with a likely range of 6 percent to 24 percent lower.[55] Because losses in soil carbon diminish over time, those researchers' findings are sensitive to the time horizon considered. They estimate that average losses in soil carbon over 10 years are 30 percent less than average losses over 5 years; other studies that have examined emissions caused by land-use-related changes in carbon dioxide emissions estimated them over a 30-year time horizon.

Critics of the new study argue that the results are based on the removal of all corn stover from a field, which is inconsistent with farmers' practices.[56] The authors, however, say that their results are insensitive to the percentage of stover that is removed. Both the critics and the authors agree that results are sensitive to a variety of soil- and crop-management practices. For example, farmers could reduce losses in soil carbon by using no-till cover crops or by applying compost or manure. Finally, as other studies have demonstrated, estimates of the emissions associated with corn stover would be lower if the lignin (the part of corn stover that typically contributes to soil carbon if the stover is left on the field) was used to generate electricity.

Switchgrass

Ethanol can also be produced from the cellulose in some perennial grasses, such as switchgrass. That type of ethanol is generally expected to result in lower emissions than corn ethanol but somewhat higher emissions than ethanol made from corn stover.

Based on a thermochemical process that it expects to be available in 2022, EPA projects that the emissions associated with switchgrass would be 72 percent lower than those associated with gasoline, with a 95 percent prediction range of 64 percent to 80 percent lower. Based on a biochemical process that it expects to be available in 2022 (and assuming that excess electricity is sold to the grid), EPA projects that switchgrass ethanol emissions would be 110 percent lower than gasoline emissions, with a 95 percent prediction range from 102 percent to 117 percent lower.

A second study, which used a simulation approach that accounted for numerous sources of uncertainty, estimated a much wider range of possible outcomes for switchgrass ethanol.[57] With a biochemical process (and electricity credits factored in), emissions from switchgrass ethanol could range from 111 percent less to 4 percent more than emissions from gasoline (based on a 95 percent prediction interval). Those authors also demonstrated the importance of electricity credits for such estimates. If the same biochemical process did not generate electricity credits, they concluded, the emissions associated with switchgrass ethanol could be as much as 99 percent higher than those associated with gasoline (their median estimate was 5 percent higher).

A third study, which used the same model and assumptions as the second study, showed how important emissions from changes in land use are to emission estimates for renewable fuels.[58] The third study concluded that if land-use-related emissions were not accounted for, the emissions from making and using switchgrass ethanol would be 16 percent to 58 percent lower than those from gasoline, even without electricity credits included, compared with a corresponding prediction range of 99 percent higher to 39 percent lower in the second study. Estimates of the emissions associated with changes in land use are very uncertain, however, which is why the 95 percent prediction ranges in the two studies differ so much.

Including electricity credits and recent estimates of landuse-related emissions (which tend to be lower than the estimates used by some earlier studies), another group of researchers estimated that emissions from switchgrass ethanol would be 77 percent to 97 percent lower than those from gasoline (based on an 80 percent prediction range).[59]

Effects of the Alternative Scenarios on Emissions

Available evidence suggests that using corn ethanol in place of gasoline has only limited potential to reduce greenhouse gas emissions (and some researchers estimate that it could actually increase emissions). Moreover, the emission estimates described above do not account for the rebound effect, in which the substitution of renewable fuels for gasoline in the United States could lead to a decrease in the price of crude oil, resulting in a rise in the global consumption of gasoline and its associated emissions. That rise in emissions elsewhere would offset some of the reduction in emissions caused by the substitution of renewable fuels for gasoline in this country. The potential of the RFS to reduce greenhouse gas emissions worldwide is generally thought to be greater to the extent that it causes gasoline and diesel to be replaced by advanced biofuels, which have relatively low emissions, rather than by corn ethanol. Such substitutions are more likely to offset any indirect increases in emissions that the RFS might trigger, including emissions resulting from changes in land use or fuel consumption elsewhere in the world.

CBO estimates that emissions in 2017 would be essentially the same whether the RFS was repealed or its requirements were kept at the proposed 2014 volumes, because those two scenarios would result in the use of very similar amounts of renewable fuels. Consumption of corn ethanol would be about the same in both scenarios, and consumption of biomass-based diesel would be, at most, not quite 2 billion gallons greater in the 2014 volumes scenario than in the repeal scenario.

Complying with the requirements stated in EISA for 2017, rather than the lower 2014 volumes, would increase the total use of renewable fuels in that year by 9 billion gallons, CBO estimates, with 7 billion of the additional gallons consisting of advanced biofuels. As a result, the potential reduction in emissions because of the RFS would be larger under that scenario than under the 2014 volumes scenario. Nonetheless, even under the EISA volumes scenario, that reduction would constitute a very small fraction of the total emissions from transportation fuels: Adding 9 billion gallons of renewable fuels to the transportation fuel supply would reduce consumption of gasoline and diesel by only about 3 percent.[60]

The effects of EISA on fuel consumption and emissions in 2017 are highly uncertain, however. For example, consumption of petroleum-based fuels could decline by more than 3 percent if increases in the prices of E10 and diesel because of the RFS led to decreases in the number of vehicle-miles traveled or to increases in the efficiency of the vehicle fleet as a whole. That decline in total fuel consumption would further decrease emissions, beyond any reduction resulting from the substitution of renewable fuels for petroleum-based fuels. Conversely, the additional increases in the use of advanced biofuels (such as sugarcane ethanol) required under the EISA volumes scenario could lead to larger increases in land-use-related emissions than have been associated with those fuels so far. That outcome would be particularly likely if farmers ran out of land with relatively small amounts of stored carbon (such as marginal farmland) and grew the additional feedstocks for those advanced biofuels on land with larger amounts of stored carbon (such as woodland).

The RFS's potential to reduce emissions will be greater over the longer term if it encourages improvements in the technology for producing fuels with relatively low emissions, such as cellulosic biofuels—particularly drop-in cellulosic gasoline or diesel that can be added to the supply of transportation fuels without encountering problems with the 10 percent blend wall. As noted above, however, tension exists between the goals of limiting the near-term cost of complying with the RFS (by issuing waivers for cellulosic biofuels, for example) and providing a strong incentive for the development of better technology.

Appendix: Details About CBO's Estimates for Food Prices and Spending

This technical appendix provides details about the estimates and assumptions that the Congressional Budget Office (CBO) used to evaluate how the Renewable Fuel Standard (RFS) might affect total U.S. spending on food in an illustrative year, 2017. The size of that effect depends on a number of factors, such as how much corn ethanol would be used in the absence of the RFS, how the production and consumption of corn respond to changes in its price, and how farmers alter their planting decisions for corn and other crops when demand for corn ethanol changes. The effect of the RFS on ethanol use is among the most important of those factors: If the RFS requirement for corn ethanol does not influence the amount of that fuel consumed in the United States, it will have no effect on food prices, regardless of other assumptions.

CBO expects that there would be little difference in the use of corn ethanol—and therefore in total spending on food—in 2017 if the requirements for renewable fuels were set at the amounts currently proposed for 2014 (the

2014 volumes scenario) or if lawmakers eliminated the RFS (the repeal scenario). Thus, this appendix focuses on the effect that the RFS would have on food spending in 2017 if the requirements for that year specified in the Energy Independence and Security Act of 2007 were met in full (the EISA volumes scenario).

Effects on Corn Ethanol Use

In CBO's estimation, the amount of corn ethanol used in 2017 would be higher under the EISA volumes scenario than under the other two scenarios by about 1.8 billion bushels, or 14 percent. In total, the RFS requires that 24 billion gallons of renewable fuels be blended into the nation's supply of gasoline and diesel in 2017. However, no more than 15 billion gallons of that requirement can be satisfied by using ethanol produced from cornstarch, leaving at least 9 billion gallons to come from other renewable fuels. CBO expects 2 billion of those 9 billion gallons to be biomass-based diesel, which means that at least 7 billion gallons would have to come from other advanced biofuels (expected to be mainly ethanol made from sugarcane and other sources of sugar).[1] Because the United States produces large quantities of corn ethanol, and because existing mandates for ethanol (which can be met with any type) are met almost entirely through the use of corn ethanol, CBO expects that under the EISA volumes scenario, fuel suppliers would use corn ethanol up to the 15 billion gallon cap specified for 2017.

Given the availability of domestically produced corn ethanol and the preponderance of its use in meeting current requirements, CBO expects that if the RFS did not exist or if its mandates were greatly reduced, fuel suppliers would want to use as much corn ethanol as possible given the constraints of the 10 percent blend wall (the fact that many older vehicles cannot operate effectively on blends of more than 10 percent ethanol). Thus, CBO projects that under either the repeal scenario or the 2014 volumes scenario, fuel suppliers would find it profitable to use 13 billion gallons of corn ethanol (an estimate based on the Energy Information Administration's forecast that U.S. consumption of blended fuel will total 132 billion gallons in 2017).[2]

Effects on the Price of Corn and on Spending for Food

The greater U.S. ethanol consumption that CBO envisions under the EISA mandates, compared with the repeal and 2014 volumes scenarios, would raise the demand for and price of corn. How much that price would increase depends on supply and demand elasticities for corn—measurements of how sensitive the production and consumption of corn are to changes in its price. An elasticity measures the percentage change in production or consumption of a good when the price changes by 1 percent.

Supply and Demand Elasticities for Corn

On the basis of various studies, CBO used values of 0.4 for the elasticity of the supply of corn and -0.25 for the elasticity of the demand for corn.[3] Those values imply that a 1 percent increase in the price of corn would raise corn production by 0.4 percent and reduce consumption by 0.25 percent. Changes to the ethanol mandate would have smaller effects on prices if supply and demand are more elastic (with elasticities farther from zero) than if they are more inelastic (with elasticities closer to zero).

More elastic conditions imply that markets can absorb changes in desired quantities bought or sold with comparatively little effect on market prices. Inelastic conditions, by contrast, imply that market prices will shift dramatically if decisions about consumption and production change, possibly even if those changes are small.

Given supply and demand elasticities, the percentage change in the market price of corn resulting from a change in the mandate for corn ethanol follows from this relationship:

$$\%\Delta P_{corn} = \frac{\%\Delta S_{corn}}{\left(\varepsilon_s - \varepsilon_d\left(\frac{S_{corn} - S^E_{corn}}{S_{corn}}\right)\right)}$$

Where S_{corn} refers to the amount of corn produced before the change in the mandate, $\%\Delta Scorn$ refers to the percentage change in the supply of corn stemming from the change in the mandate, ε_s and ε_d refer to the supply and demand elasticities for corn, and S^E_{corn} refers to the amount of corn used to produce ethanol before the change in the mandate.[4] The last term in the denominator,

$$\frac{S_{corn} - S^{E}_{corn}}{S_{corn}}$$

denotes the percentage of U.S. corn production that is used for purposes other than making ethanol.

Changes in U.S. Ethanol Production and Their Effect on the Demand for Corn

On the basis of its most recent agricultural outlook, CBO estimates that the United States would produce about 14 billion bushels of corn in 2017 under the 2014 volumes scenario or the repeal scenario, with about 5 billion of those bushels used to make ethanol.[5] The 14 percent higher U.S. ethanol production assumed to occur under the EISA volumes scenario would raise the demand for corn for ethanol production by about 0.7 billion bushels.

Table A-1. Production and Price of Corn in 2017 Under CBO's Alternative Scenarios for the Renewable Fuel Standard

	2014 Volumes Scenario and Repeal Scenario[a]	EISA Volumes Scenario[b]	Percentage Difference Between the EISA Volumes Scenario and the Other Two Scenarios
Projected U.S. Corn Production in 2017 (Billions of bushels)	14.1	14.4	2
Projected Average Price of Corn in 2017 (Dollars per bushel)	4.25	4.50	6[c]

Source: Congressional Budget Office.

[a] For the 2014 volumes scenario, CBO assumed that the 2017 requirements for renewable fuels would be set at the same volumes that the Environmental Protection Agency (EPA) has proposed for 2014. For the repeal scenario, CBO

assumed that lawmakers would repeal the Renewable Fuel Standard in 2014, so fuel suppliers would not be subject to any requirements for the use of renewable fuels in 2017. CBO projects that total use of corn ethanol in the United States would be about the same under both scenarios.

[b] For this scenario, CBO assumed that fuel suppliers would have to comply with the various requirements for renewable fuels that are specified for 2017 in the Energy Independence and Security Act of 2007 (except the requirement for cellulosic biofuels, which EPA is assumed to reduce to a volume consistent with the projected production capacity for those fuels).

[c] This calculation is based on assumed elasticities of 0.4 for the supply of corn and -0.25 for the demand for corn and on CBO's projection that 5 billion bushels of corn will be used to produce ethanol in 2017. Those elasticities imply that a 1 percent increase in the price of corn would raise corn production by 0.4 percent and reduce corn consumption by 0.25 percent. (For more about those elasticities, see the text.)

The overall demand for corn under that scenario would increase by a smaller amount, however. The reason is that ethanol production creates byproducts that are used as animal feed (1 bushel of corn used in ethanol production produces roughly 0.3 bushels of animal feed).[6] With more of those byproducts available, the amount of corn used as animal feed would decline somewhat. Accounting for those byproducts means that the net change in total demand for corn under the EISA volumes scenario (in the absence of any other changes in the quantity demanded once the market price rises) would be an increase of about 0.4 billion bushels—or roughly 3 percent of the U.S. corn production that would occur in 2017 without the RFS, CBO estimates. With changes in demand caused by higher corn prices factored in, the net increase in corn production would be smaller, about 0.3 billion bushels, or 2 percent.

Effects on the Price of Corn

With the elasticities for corn used in this analysis, CBO estimates that the 14 percent higher production of corn ethanol under the EISA volumes scenario than under the other scenarios means that the price of corn would be about 6 percent higher under the EISA volumes scenario. CBO projects that the price of corn will average about $4.25 per bushel in 2017 under the 2014 volumes scenario or the repeal scenario, so that 6 percent increase would raise the price per bushel by about 25 cents (see Table A-1).

Effects on the Cost of Food

The estimated increase in the price of corn would cause total U.S. spending on food in 2017 to be $3.5 billion higher under the EISA volumes scenario than under the other two scenarios, CBO estimates. That rise represents about one-quarter of one percent of the roughly $1.8 trillion expected to be spent on food in the United States during 2017 under the repeal scenario or the 2014 volumes scenario. CBO's estimate is based on the ways in which a higher corn price would affect the costs of food products that contain corn, food products from animals that eat corn, and agricultural products whose production would be displaced by corn.

Table A-2. Effects of the EISA Volumes Scenario on Food Prices and Spending in 2017

	Change Under the EISA Volumes Scenario
Estimated Change in the Average Price of Corn (Dollars per bushel)	0.25
Effects of the Higher Corn Price on Food Prices	
Direct effects through the cost of corn products	
Total amount of corn used in food in 2017 (Billions of bushels)	1.5
Change in spending on food products that contain corn (Billions of dollars)	0.4
Indirect effects through the cost of meat, poultry, and dairy products	
Total amount of corn used in animal feed in 2017 (Billions of bushels)	5.4
Change in spending on meat, poultry, and dairy products (Billions of dollars)	1.3
Indirect effects through the cost of soybean products	
Change in U.S. corn production (Billions of bushels)	0.3
Average corn yield (Bushels per acre)	170
Change in total acres planted in corn (Millions)	1.9
Change in U.S. soybean production (Percent)	-2
Change in the average price of soybeans (Dollars per bushel)	1.00
Total U.S. soybean production (Billions of bushels)	1.8

	Change Under the EISA Volumes Scenario
Change in spending on soybeans (Billions of dollars)	1.8
Total Change in Spending on Food (Billions of dollars)	3.5
Projected 2017 Food Expenditures (Billions of dollars)	1,770
Percentage Change in Spending on Food	0.2

Source: Congressional Budget Office.

Notes: For the EISA volumes scenario, CBO assumed that fuel suppliers would have to comply with the various requirements for renewable fuels that are specified for 2017 in the Energy Independence and Security Act of 2007 (except the requirement for cellulosic biofuels, which the Environmental Protection Agency is assumed to reduce to a volume consistent with the projected production capacity for those fuels).

These calculations are based on the effect of the EISA volumes scenario on the average price of corn in 2017 as shown in Table A-1. The calculations assume that increases in the prices of corn and soybeans are passed on fully to consumers in the form of higher prices for food.

Spending on Corn Products

CBO expects that roughly 1.5 billion of the 14 billion bushels of corn produced in the United States in 2017 under either the 2014 volumes scenario or the repeal scenario would be used directly in food. That quantity includes corn consumed on its own, corn used in breakfast cereals and other food products, and corn used to produce corn syrup and other sweeteners. With the price of corn about 25 cents higher per bushel under the EISA volumes scenario, direct spending on corn products in 2017 would be about $0.4 billion higher than under the other scenarios, if the higher corn price was fully passed on to consumers in the form of higher prices for food (see Table A-2). The difference in overall spending on corn products would be smaller if some of the higher cost of corn was absorbed by intermediaries at various stages of production, transportation, packaging, or marketing.

Spending on Meat, Poultry, and Dairy Products

CBO expects 5.4 billion bushels of corn to be used as animal feed in 2017 under the 2014 volumes scenario or the repeal scenario.[7] Because more corn is used for animal feed than is used directly for food production, higher costs for meat, poultry, and dairy products would probably have a larger effect on food prices overall than would higher costs for corn products. If farmers passed the full increase in animal feed prices along to retail consumers, the higher price of corn in the EISA volumes scenario would increase consumers' total spending on meat, poultry, and dairy products in 2017 by about $1.3 billion, CBO estimates. As with consumers' spending on corn products, the rise in spending would be smaller if intermediaries absorbed some of the cost increases as feed prices rose.

Spending on Other Agricultural Products

Higher demand for corn can also affect the price of food indirectly as farmers increase the amount of land they use to grow corn and reduce the amount they use for other crops, thus raising the prices of those other crops. How many acres would be converted under the EISA volumes scenario depends on how much more corn would be produced domestically and on the average yield per acre of corn. CBO expects corn yields to average about 170 bushels per acre in 2017.[8] Producing 1.8 billion more gallons of corn ethanol and 0.3 billion more bushels of corn (the net effect of more corn used for ethanol production and less used for other purposes because of higher market prices) would require almost 2 million more acres of U.S. farmland to be used for corn production in the EISA volumes scenario than in the other two scenarios.

Much of that land would formerly have been used for growing other crops. A lesser supply of those crops would increase their prices, further pushing up the cost of food. After corn, soybeans and wheat are the two crops grown in the greatest amounts in the United States.[9] Thus, most of the additional acreage that would be used for corn under the EISA volumes scenario would otherwise have been used for growing those crops. Corn and soybeans are typically planted in rotation (with a crop of one followed by a crop of the other on the same land), so CBO expects that all of the additional land that would be used to grow corn under the EISA volumes scenario in 2017 would previously have been used to grow soybeans. (If some of that available land had been used to produce other crops, CBO's conclusions would not be affected substantially.)

Using almost 2 million fewer acres to grow soybeans would reduce projected soybean production by about 2 percent. CBO estimates that the price elasticity of demand for soybeans is the same as for corn, -0.25.[10] Thus, a 2 percent reduction in soybean production would raise the price of soybeans by about 10 percent—or $1.00 per bushel, based on CBO's projection of the price of soybeans in 2017 under the repeal scenario or the 2014 volumes scenario. With 1.8 billion bushels of soybeans expected to be harvested in that year under either the repeal scenario or the 2014 volumes scenario, CBO estimates that the higher soybean price under the EISA volumes scenario would add $1.8 billion to total spending on food in 2017, if the price increase is passed on fully to consumers in the form of higher food costs.

End Notes

[1] Unless otherwise indicated, the amounts of biomass-based diesel discussed in this report are measured in "compliance-equivalent gallons." Under EISA, 1 gallon of biomass-based diesel is considered equivalent to 1.5 gallons of ethanol for purposes of complying with the RFS.

[2] Cellulosic feedstocks can be used to make diesel or gasoline as well as to make biofuels. A gallon of cellulosic diesel would count toward satisfying either the cellulosic biofuel mandate or the biomass-based diesel mandate.

[3] If a fuel supplier that is obligated to meet the RFS is out of compliance at the end of a year (after accounting for its RINs and its use of renewable fuels), EPA may fine the supplier as much as $32,500 per day, plus the savings to the supplier that result from its noncompliance. Those penalties are specified in sections 205 and 211(d) of the Clean Air Act, 42 U.S.C. §§7524, 7545(d) (2012).

[4] See Energy Information Administration, *Annual Energy Outlook 2013, With Projections to 2040*, DOE/EIA-0383(2013) (April 2013), Figure 100, www.eia.gov/forecasts/archive /aeo13.

[5] Flex-fuel vehicles are identical to ordinary passenger vehicles except for slight differences in their fuel systems and, in many cases, an identifying badge on a fender or rear panel. According to the Department of Energy, many owners of flex-fuel vehicles are not aware that their vehicles can run on blends of more than 10 percent ethanol. See Department of Energy, "Alternative Fuels Data Center—Flexible Fuel Vehicles" (October 3, 2013), www.afdc.energy

[6] See Kristi Moriarty, "E85 Deployment" (presentation prepared for the Energy Information Administration's biofuels workshop by the National Renewable Energy Laboratory, March 20, 2013), www.eia.gov/biofuels/workshop/presentations/2013.

[7] Industry groups challenged EPA's certification of E15 in court. The Supreme Court recently dismissed those challenges, which prompted the Alliance of Automobile Manufacturers to assert that "vehicles [built since 2001] were never designed to run on this more corrosive fuel. Automakers continue to urge consumers to check their owner's manuals for the recommended fuel to use safely in their vehicles." See Alliance of Automobile Manufacturers, "Alliance Response to Supreme Court Decision Today to Dismiss Challenges to EPA's E15 Decision" (June 24, 2013), http://tinyurl.com/q8um8eg.

[8] See Joanna Schroeder, "EPA Announces 2014 RVO Numbers for RFS" (November 15, 2013), http://tinyurl.com/oh9vfo4.

[9] For the legal decision about the 2012 mandate, see *API v. EPA*, 706 F.3d 474 (D.C. Cir. 2013).

[10] See Environmental Protection Agency, *EPA Issues Direct Final Rule for 2013 Cellulosic Standard*, EPA-420-F-14-018 (April 2014), http://go.usa.gov/9rd3 (PDF, 151 KB).

[11] EPA has not relieved suppliers of their compliance obligations for those years but instead has allowed them to satisfy the obligations in a different way. Specifically, whenever EPA has reduced the RFS mandate on cellulosic biofuels, it has offered credits for sale to fuel suppliers in an amount equal to the new, revised mandate. If suppliers plan to substitute some other advanced biofuel for cellulosic biofuel, they must buy a waiver credit from EPA as well as the gallon of that other fuel. EPA determines the price of waiver credits on the basis of the previous year's wholesale price of gasoline.

[12] See Environmental Protection Agency, *EPA Proposes 2014 Renewable Fuel Standards, 2015 Biomass-Based Diesel Volume,* EPA-420- F-13-048 (November 2013), Table 1, p. 3, http://go.usa.gov/9rdA (PDF, 190 KB). Although EPA announced its proposal for the 2014 requirement in November 2013, it has not yet issued a final rule. The delay may reflect uncertainty because of ongoing legal challenges to its revisions to the 2013 requirement, which it finalized in August 2013.

[13] EISA allows EPA to set the requirement for biomass-based diesel at a volume not lower than 1 billion gallons. Each gallon of biomass-based diesel provides 1.5 RINs for the purposes of complying with the advanced-biofuel requirement, so the requirement for 1 billion gallons accounts for 1.5 billion gallons of compliance. EPA has not yet set that volume for 2017. For illustrative purposes, CBO assumed that it would be 2 billion compliance-equivalent gallons, only slightly more than the 2014 requirement.

[14] In calculating that gap, CBO accounted for the fact that each gallon of biomass-based diesel provides 1.5 RINs for the purposes of complying with the advanced-biofuel requirement. So far, the largest gap occurred in 2012, when the requirement for advanced biofuels was set at 2 billion gallons and the requirement for biomass-based diesel was set at 1 billion gallons (1.5 billion on a compliance-equivalent basis).

[15] See Energy Information Administration, *Annual Energy Outlook 2014, With Projections to 2040*, DOE/EIA-0383(2014) (April 2014), Figure MT-56, www.eia.gov/forecasts/aeo.

[16] That figure is CBO's estimate based on Energy Information Administration, *International Energy Outlook 2013, With Projections to 2040,* DOE/EIA-0484(2013) (July 2013), www.eia.gov/forecasts/ieo.

[17] Some industry observers have speculated that larger U.S. imports of sugarcane ethanol from Brazil could be achieved not by encouraging increased production in Brazil but by exchanging sugarcane ethanol made in that country for corn ethanol made in the United States. (That type of swap has already taken place to a limited degree, as discussed in Energy Information Administration, *Biofuels Issues and Trends*, October 2012, www.eia.gov/biofuels/ issuestrends.) CBO did not estimate the cost of that approach because it would require Brazil to replace 60 percent of its consumption of domestically produced sugarcane ethanol with corn ethanol imported from the United States. Besides posing large logistical challenges, such an exchange would not increase the global use of advanced biofuels but would consume scarce resources and produce additional greenhouse gas emissions to transport the swapped ethanol supplies.

[18] See Scott Irwin and Darrel Good, "Recent Trends in Biodiesel Prices and Production Profits" (Department of Agricultural and Consumer Economics, University of Illinois Urbana-Champaign, September 18, 2013), http://farmdocdaily.illinois.edu/2013/09; and Don Hofstrand, "Tracking Biodiesel Profitability" (Iowa State University Extension and Outreach, July 2012), www.extension.iastate.edu/agdm/energy

[19] See Ralph Groschen, *Overview of the Feasibility of Biodiesel From Waste/Recycled Greases and Animal Fats* (Minnesota Department of Agriculture, October 2002), http://tinyurl.com/l6222gb (PDF, 319 KB).

[20] For expectations that ethanol will cost less per gallon than gasoline in 2017, see CME Group, "RBOB Gasoline Futures" and "CBOT Denatured Fuel Ethanol Futures" (accessed June 23, 2014), www.cmegroup.com. The price of ethanol is about the same as that of gasoline per British thermal unit (Btu) of energy content; however, analysts generally believe that for blends of 10 percent ethanol or less, fuel suppliers make choices based on the per-gallon cost of the two fuels rather than the per-Btu cost. See Scott Irwin and Darrel Good, "Ethanol Blending Margins, RFS2 Compliance, and the Price of Gasoline" (AgFax, April 3, 2012), http://tinyurl.com/qj6ltg7.

[21] For an earlier study on that topic, see Congressional Budget Office, *The Impact of Ethanol Use on Food Prices and Greenhouse Gas Emissions* (April 2009), www.cbo.gov/publication/41173.

[22] An increase in the prices of certain types of food would cause consumers to reduce the amount of those foods that they purchased. But because the effects on food prices in this analysis are small, any reduction in the amounts of certain types of food consumed would also be small, and it would be offset at least in part by increased consumption of other types of food. Thus, CBO's calculations reflect the assumption that the total quantity of food purchased would be unaffected by the increase in food prices.

[23] Some studies have estimated that the RFS would result in only a small increase, or even a decrease, in the price of E10. See, for example, Antonio M. Bento, Richard Klotz, and Joel Landry, "Are There Carbon Savings from U.S. Biofuel Policies? The Critical Importance of Accounting for Leakage in Land and Fuel Markets," *Energy Journal* (forthcoming), http://dx.doi.org/10.2139/ ssrn.2219503; and Deepak Rajagopal and Richard J. Plevin, "Implications of Market-Mediated Emissions and Uncertainty for Biofuel Policies," *Energy Policy,* vol. 56 (May 2013), pp. 75–82, http://dx.doi.org/10.1016/j.enpol.2012.09.076. Those studies reached different conclusions than CBO did for several reasons, including the following: They allowed a longer time (10 to 15 years) for the market to adjust to a required increase in the use of renewable fuels; they assumed that gasoline could absorb an ethanol concentration of up to 20 percent without encountering problems with the availability of filling stations or the demand for that blend of fuel (although such problems would probably arise); and they did not model all of the compliance requirements of the RFS (for example, they did not account for the fact that suppliers of E10 would need to buy biomass-based diesel RINs).

[24] The price of renewable fuel RINs rose in 2013—briefly spiking above $1 in the summer of that year—because market participants were buying RINs and saving them to use in 2014 in the expectation that the RFS would push the required use of ethanol beyond the 10 percent blend wall in 2014 (as would happen if EPA set the mandates at the 2014 volumes stated in EISA). Fuel suppliers responded to that rise in part by buying additional biomass-based diesel RINs. When it seemed likely that EPA would modify the renewable fuel requirement so that the blend wall would not be exceeded, the price of renewable fuel RINs dropped. See Scott Irwin, "Will the EPA Reverse Itself on the Write Down of the Renewable Mandate for 2014? The Message From the RINs Market" (Department of Agricultural and Consumer Economics, University of Illinois Urbana-Champaign, February 19, 2014), http://farmdocdaily.illinois.edu/2014/02.

[25] For a discussion of the 10 cent per gallon estimate, see Bruce A. Babcock and Sebastien Pouliot, *Impact of Sales Constraints and Entry on E85 Demand*, Policy Brief 13-PB 12 (Center for Agricultural and Rural Development, Iowa State University, August 2013), http://tinyurl.com/k6df9lu.

[26] That estimate of 17 million is based on the current number of flex-fuel vehicles on the road and the average annual net increase in their numbers in recent years.

[27] See Bruce A. Babcock and Sebastian Pouliot, *Price It and They Will Buy: How E85 Can Break the Blend Wall*, Policy Brief 13-PB 11 (Center for Agricultural and Rural Development, Iowa State University, August 2013), http://tinyurl.com/nodq3k5.

[28] That range is based on a study in which researchers estimate that a 40 percent cost savings may be necessary to cause the market to accommodate an additional 6 billion gallons of ethanol by 2015. Their calculation is based on the assumptions that vehicle owners will drive up to 10 miles out of their way to obtain E85 and that existing E85 pumps can deliver as much additional fuel as needed to meet local demand. See Bruce A. Babcock and Sebastian Pouliot, *Price It and They Will Buy: How E85 Can Break the Blend Wall*, Policy Brief 13-PB 11 (Center for Agricultural and Rural Development, Iowa State University, August 2013), http://tinyurl.com/nodq3k5. Those authors' subsequent research concluded that existing pumps would be insufficient to supply an additional 6 billion gallons. However, given the longer time frame available for compliance in this analysis (2017 rather than 2015), CBO expects that sufficient pumps could become available by 2017.

[29] That calculation is based on the Energy Information Administration's projection that the wholesale price of E10 will be about $3 per gallon in 2017 and CBO's estimate that corn prices will average about $4.80 per bushel in 2017 if the corn ethanol mandate is met in full. See Energy Information Administration, *Annual Energy Outlook 2014, With Projections to 2040*, DOE/ EIA-0383(2014) (April 2014), www.eia.gov/forecasts/aeo. The calculation of the RIN price necessary to generate the required level of savings follows the approach used in Bruce A. Babcock and Sebastian Pouliot, *Price It and They Will Buy: How E85 Can Break the Blend Wall*, Policy Brief 13-PB 11 (Center for Agricultural and Rural Development, Iowa State University, August 2013), http://tinyurl.com/nodq3k5.

[30] On the basis of its most recent agricultural outlook (summarized in Congressional Budget Office, "USDA Mandatory Farm Programs—April 2014 Baseline," www.cbo.gov/publication/44202), CBO estimates that consuming roughly 2 billion additional gallons of corn ethanol would raise the average price of corn in 2017 by about 25 cents per bushel (as explained in the appendix). A bushel of corn yields about 2.5 gallons of ethanol, so the 25 cent rise in the price of corn—if entirely passed on to consumers in the form of higher prices for finished products—would increase the price of corn ethanol by about 10 cents per gallon.

[31] Brazil exported about 2.9 million cubic meters (750 million gallons) of ethanol in 2013. The United States, the European Union, and Caribbean countries received the largest shares of those exports. See Brazilian Sugarcane Industry Association, "Monthly Report of Brazilian Ethanol Exports, Calendar Year 2014," http://tinyurl.com/ktawewp (accessed April 24, 2014).

[32] See Gal Hochman, Deepak Rajagopal, and David Zilberman, "The Effect of Biofuels on the International Oil Market," *Applied Economic Perspectives and Policy*, vol. 33, no. 3 (Autumn 2011), pp. 402–407, http://dx.doi.org/10.1093/aepp/ppr016; and Congressional Budget Office, *Energy Security in the United States* (May 2012), www.cbo.gov/publication/43012.

[33] The percentage increases in cost are based on the Energy Information Administration's most recent forecast for the retail price of diesel in 2017. See Energy Information Administration, *Annual Energy Outlook 2014, With Projections to 2040,* DOE/ EIA-0383(2014) (April 2014), Table A12, "Petroleum and Other Liquids Prices," www.eia.gov/forecasts/aeo.

[34] That value includes the additional 10 cents per gallon that the fuel supplier would pay for each gallon of corn ethanol. Thus, the supplier would receive $1.45 to $2.00 of net revenue for any RIN that it sold. Likewise, the opportunity cost if the supplier used a RIN for its own compliance purposes would be $1.45 to $2.00.

[35] That example uses E10 made with corn ethanol; however, the price increase for E10 would not depend on whether corn ethanol or sugarcane ethanol was used. Instead, the cost of complying with the RFS would depend on the quantities of corn ethanol and sugarcane ethanol that the standard required. That compliance cost, with the estimates used in this analysis, would be $13 to $26 for each 100 gallons of E10.

[36] The percentage increases in cost are based on the Energy Information Administration's most recent forecast for the retail price of blended gasoline in 2017. See Energy Information Administration, *Annual Energy Outlook 2014, With Projections to 2040,* DOE/EIA-0383(2014) (April 2014), Table A12, "Petroleum and Other Liquids Prices," www.eia.gov/forecasts/aeo.

[37] Ibid., Table 11.

[38] The percentage increases in cost are based on the Energy Information Administration's most recent forecast for the retail price of E85 in 2017. See Energy Information Administration, *Annual Energy Outlook 2014, With Projections to 2040,* DOE/EIA-0383(2014) (April 2014), Table A12, "Petroleum and Other Liquids Prices," www.eia.gov/forecasts/aeo.

[39] For example, carbon dioxide is released when soil is disturbed, when trees are burned, or when the residue left from harvests decays. Such changes typically cause a large, one-time release of greenhouse gases initially, which means that producing and consuming ethanol can result in a very different time profile of emissions than producing and consuming gasoline. To account for that difference, researchers typically compare the emissions from ethanol and gasoline over an extended period (generally 30 years). Estimating emission reductions over longer periods allows more years for a one-time release of carbon dioxide to be offset by later emission reductions that occur through production and consumption.

4[0] If everything else is equal, a lower world price of oil will reduce the price of gasoline in the United States. The effect of the RFS on the price of blended fuel could be positive or negative, however, depending on the relative magnitude of the decrease in the wholesale price of gasoline and the increase in the marginal cost of producing renewable fuels.

[41] See Gal Hochman, Deepak Rajagopal, and David Zilberman, "The Effect of Biofuels on the International Oil Market," *Applied Economic Perspectives and Policy*, vol. 33, no. 3 (Autumn 2011), pp. 402–407, http://dx.doi.org/10.1093/aepp/ppr016.

[42] Most of the studies reviewed estimated a range of possible outcomes. The 29 percent to 85 percent span represents the highest and lowest estimates from those ranges. See Deepak Rajagopal, "The Fuel Market Effects of Biofuel Policies and Implications for Regulations Based on Lifecycle Emissions," *Environmental Research Letters*, vol. 8, no. 2 (April 2013), http://dx.doi.org/10.1088/1748-9326/8/2/024013.

[43] Those estimates take into account differences in the energy content of renewable fuels and petroleum-based fuels. Because renewable fuels have lower energy content than petroleum-based gasoline or diesel, each gallon of renewable fuel replaces less than one full gallon of gasoline or diesel.

[44] See Environmental Protection Agency, *Renewable Fuel Standard Program (RFS2) Regulatory Impact Analysis*, EPA-420-R-10-006 (February 2010), www.epa.gov/otaq/renewablefuels/420r10006.pdf (17 MB).

[45] See National Research Council, *Renewable Fuel Standard: Potential Economic and Environmental Effects of U.S. Biofuel Policy* (National Academies Press, 2011),

http://tinyurl.com/89yfx36. Other researchers say that an advantage of EPA's approach is that it uses a unified modeling framework to determine the changes in emissions associated with production and consumption of renewable fuels and food crops, capturing potential market interactions that might otherwise be missed. In contrast, most other estimates of net emissions are obtained by combining independently derived estimates of direct reductions and indirect effects. For more discussion, see Deepak Rajagopal, "Consequential Life Cycle Assessment of Policy Vulnerability to Price Effects," *Journal of Industrial Ecology*, vol. 18, no. 2 (April 2014), pp. 164–175, http://dx.doi.org/10.1111/jiec.12058; and Richard J. Plevin, Mark A. Delucchi, and Felix Creutzig, "Using Attributional Life Cycle Assessment to Estimate Climate Change Mitigation Benefits Misleads Policymakers," *Journal of Industrial Ecology*, vol. 18, no. 1 (February 2014), pp. 73–83, http://dx.doi.org/10.1111/jiec.12074.

[46] See National Research Council, *Renewable Fuel Standard: Potential Economic and Environmental Effects of U.S. Biofuel Policy* (National Academies Press, 2011), http://tinyurl.com/89yfx36. One of the studies estimated a 90 percent increase in emissions for each gallon of ethanol used in place of gasoline; however, that study relied on estimates of land-use-related emissions that have been criticized in recent years. For a discussion of those criticisms, see Roger A. Sedjo, Brent Sohngen, and Anne Riddle, *Wood Bioenergy and Land Use: A Challenge to the Searchinger Hypothesis*, Discussion Paper RFF DP 13-33 (Resources for the Future, November 2013), www.rff.org/RFF/Documents/RFF-DP-13-33.pdf (711 KB).

[47] See Kimberley A. Mullins, W. Michael Griffin, and H. Scott Matthews, "Policy Implications of Uncertainty in Modeled Life-Cycle Greenhouse Gas Emissions of Biofuels," *Environmental Science and Technology*, vol. 45, no. 1 (January 1, 2011), pp. 132– 138, http://dx.doi.org/10.1021/es1024993. The study measured emissions over a 30-year period and did not discount future emission reductions.

[48] See Michael Q. Wang and others, "Well-to-Wheels Energy Use and Greenhouse Gas Emissions of Ethanol From Corn, Sugarcane, and Cellulosic Biomass for U.S. Use," *Environmental Research Letters*, vol. 7, no. 4 (October–December 2012), http://dx.doi.org/10.1088/1748-9326/7/4/045905.

[49] See Michael Q. Wang and others, "Energy and Greenhouse Gas Emission Effects of Corn and Cellulosic Ethanol With Technology Improvements and Land Use Changes," *Biomass and Bioenergy*, vol. 35, no. 5 (May 2011), pp. 1885–1896, http://dx.doi.org/10.1016/j.biombioe.2011.01.028; and Jennifer B. Dunn and others, "Land-Use Change and Greenhouse Gas Emissions From Corn and Cellulosic Ethanol," *Biotechnology for Biofuels*, vol. 6, no. 51 (April 2013), http://dx.doi.org/10.1186/ 1754-6834-6-51.

[50] See Environmental Protection Agency, *Renewable Fuel Standard (RFS2) Regulatory Impact Analysis*, EPA-420-R-10-006 (February 2010), p. 337, www.epa.gov/otaq/renewablefuels/420r10006.pdf (17 MB).

[51] See Michael Q. Wang and others, "Well-to-Wheels Energy Use and Greenhouse Gas Emissions of Ethanol From Corn, Sugarcane, and Cellulosic Biomass for U.S. Use," *Environmental Research Letters*, vol. 7, no. 4 (October–December 2012), http://dx.doi.org/10.1088/1748-9326/7/4/045905.

[52] Based on total use in 2013 as reported in Energy Information Administration, "Monthly Biodiesel Production Report" (January 2014), www.eia.gov/biofuels/biodiesel

[53] See Michael Q. Wang and others, "Well-to-Wheels Energy Use and Greenhouse Gas Emissions of Ethanol From Corn, Sugarcane, and Cellulosic Biomass for U.S. Use,"

Environmental Research Letters, vol. 7, no. 4 (October–December 2012), http://dx.doi.org/10.1088/1748-9326/7/4/045905.

[54] See Sabrina Spatari and Heather L. MacLean, "Characterizing Model Uncertainties in the Life Cycle of Lignocellulose-Based Ethanol Fuels," *Environmental Science and Technology,* vol. 44, no. 22 (November 15, 2010), pp. 8773–8780, http://dx.doi.org/ 10.1021/es102091a. Calculations are based on the 50th percentile outcome in Figure 1 of that study.

[55] See Adam J. Liska and others, "Biofuels From Crop Residue Can Reduce Soil Carbon and Increase CO_2 Emissions," *Nature Climate Change*, vol. 4 (April 20, 2014), http://dx.doi.org/ 10.1038/nclimate2187. That range represents plus or minus two standard deviations from the mean estimate of 15 percent; according to that study, actual emission reductions would be likely to lie within that range.

[56] See, for example, National Corn Growers Association, "Federal Agencies, Others Dispute Report on Corn Stover Ethanol and Climate Change," *National Journal* (April 23, 2014), www.nationaljournal.com/library/143649; and Donnelle Eller, "Experts Say Ethanol Study Used Bad Model," *Des Moines Register* (April 21, 2014), http://tinyurl.com/qfppjyj.

[57] See Sabrina Spatari and Heather L. MacLean, "Characterizing Model Uncertainties in the Life Cycle of Lignocellulose-Based Ethanol Fuels," *Environmental Science and Technology*, vol. 44, no. 22 (November 15, 2010), pp. 8773–8780, http://dx.doi.org/ 10.1021/es102091a.

[58] See Yimin Zhang, Sabrina Spatari, and Garvin Heath, "Are We Ready for Consequential LCA-Based Regulations?" (presentation prepared by the National Renewable Energy Laboratory for the American Center for Life Cycle Assessment's LCA X Conference, Portland, Oregon, November 2, 2010), www.lcacenter.org/ LCAX/presentations-final/145.pdf (746 KB).

[59] See Michael Q. Wang and others, "Well-to-Wheels Energy Use and Greenhouse Gas Emissions of Ethanol From Corn, Sugarcane, and Cellulosic Biomass for U.S. Use," *Environmental Research Letters*, vol. 7, no. 4 (October–December 2012), http://dx.doi.org/10.1088/1748-9326/7/4/045905.

[60] That calculation is based on the Energy Information Administration's projections of gasoline and diesel consumption in 2017 and on the assumptions that renewable fuels have two-thirds of the energy content of petroleum fuels and that the RFS does not alter the total energy content of transportation fuels.

End Notes for Appendix I

[1] The RFS requires at least 1 billion gallons of biomass-based diesel be used to meet the 2017 mandate, although the exact amount will depend on future rulemaking. One gallon of biomass-based diesel can be used in place of 1.5 gallons of ethanol for purposes of complying with the RFS, which is why CBO expects that about 2 billion gallons of biomass-based diesel will be used in 2017.

[2] See Energy Information Administration, *Annual Energy Outlook 2014, With Projections to 2040,* DOE/EIA-0383(2014) (April 2014), www.eia.gov/forecasts/aeo.

[3] Estimates of elasticity were taken from Harry de Gorter and David R. Just, *The Welfare Economics of an Excise-Tax Exemption for Biofuels and the Interaction Effects With Farm Subsidies*, Working Paper 2007-13 (Cornell University, Department of Applied Economics and Management, December 2007), http://tinyurl.com/poyu732 (PDF, 427 KB); Bruce Gardner, "Fuel Ethanol Subsidies and Farm Price Support," *Journal of Agricultural and*

Food Industrial Organization, vol. 5, no. 2 (December 2007), http://dx. doi. org/10.2202/1542-0485.1188; Andrew Schmitz, Charles B. Moss, and Troy G. Schmitz, "Ethanol: No Free Lunch," *Journal of Agricultural and Food Industrial Organization*, vol. 5, no. 2 (December 2007), http://dx.doi.org/10.2202/1542-0485.1186; Scott Baier and others, *Biofuels Impact on Crop and Food Prices: Using an Interactive Spreadsheet*, International Finance Discussion Papers 967 (Board of Governors of the Federal Reserve System, March 2009), www.federalreserve.gov/pubs/ifdp/2009/967/ default.htm; and Paul W. Gallagher and others, "Some Long-Run Effects of Growing Markets and Renewable Fuel Standards on Additives Markets and the U.S. Ethanol Industry," *Journal of Policy Modeling,* vol. 25, no. 6–7 (September 2003), pp. 585–608, http://dx. doi. org/10.1016/S0161-8938 (03) 00055-3.

[4] See Scott Baier and others, *Biofuels Impact on Crop and Food Prices: Using an Interactive Spreadsheet,* International Finance Discussion Papers 967 (Board of Governors of the Federal Reserve System, March 2009), www.federalreserve.gov/pubs/ifdp/2009/ 967/default.htm.

[5] That outlook is summarized in Congressional Budget Office, "USDA Mandatory Farm Programs—April 2014 Baseline" (April 2014), www.cbo.gov/publication/44202.

[6] See Dusan Drabik, *The Theory of Biofuel Policy and Food Grain Prices*, Working Paper 2011-20 (Cornell University, Charles H. Dyson School of Applied Economics and Management, March 2012), http://tinyurl.com/q2kj54g (PDF, 580 KB).

[7] See Congressional Budget Office, "USDA Mandatory Farm Programs—April 2014 Baseline" (April 2014), www.cbo.gov/ publication/44202.

[8] See Congressional Budget Office, "USDA Mandatory Farm Programs—April 2014 Baseline" (April 2014), www.cbo.gov/ publication/44202.

[9] See Department of Agriculture, National Agricultural Statistics Service, *Crop Production* (September 12, 2013), http://tinyurl.com/oavucc5 (PDF, 792 KB).

[10] Estimated elasticities for soybeans were taken from Scott Baier and others, *Biofuels Impact on Crop and Food Prices: Using an Interactive Spreadsheet*, International Finance Discussion Papers 967 (Board of Governors of the Federal Reserve System, March 2009), www.federalreserve.gov/pubs/ifdp/2009/967/default.htm; and William Lin and others, *Supply Response Under the 1996 Farm Act and Implications for the U.S. Field Crops Sector*, Technical Bulletin TB-1888 (Department of Agriculture, Economic Research Service, September 2000), http://go.usa.gov/kkVz.

INDEX

C

D

E

F

L

M

N

S

T